M. Rossbach, J.D. Schladot,
P. Ostapczuk (Eds.)

Specimen Banking

Environmental Monitoring and Modern Analytical Approaches

With 65 Figures

Springer-Verlag
Berlin Heidelberg New York
London Paris Tokyo
Hong Kong Barcelona
Budapest

Editors:

Dr. M. Rossbach,
Dr. J.D. Schladot,
Dr. P. Ostapczuk

Forschungszentrum Jülich GmbH, Institut für,
Angewandte Physikalische Chemie, Umweltprobenbank,
Postfach 1913, 5170 Jülich, FRG

ISBN-13: 978-3-642-77199-6   e-ISBN-13: 978-3-642-77197-2
DOI: 10.1007/978-3-642-77197-2

Typesetting: Macmillan India Ltd, Bangalore-25;

51/3020 5 4 3 2 1 0 – Printed on acid-free paper

List of Contributors

Ambe, Y., Prof. Dr.
Faculty of Agriculture, Tokyo University of Agriculture and Technology,
Fuchu, Tokyo 183, Japan

Backhaus, F.W., Dipl.-Ing.
Kernforschungsanlage Jülich, Institut für Angewandte Physikalische Chemie,
Postfach 1913, 5170 Jülich, FRG

Bertram, H.P., Prof. Dr.
Institut für Pharmakologie und Toxikologie der Universität Münster,
Umweltprobenbank, Domagkstraße 12, 4400 Münster/Westfalen, FRG

Boehringer, U.R., Dr.
Umweltbundesamt, Bismarckplatz 1, 1000 Berlin 33, FRG

Dürbeck, H.W., Dr. Dr. Habil.
Institut für Angewandte, Physikalische Chemie, Forschungszentrum Jülich,
Postfach 1913, 5170 Jülich, FRG

Erdtmann, G., Dr.
Zentralabteilung für Chemische Analyse, ZCH, Forschungszentrum Jülich,
KFA, 5170 Jülich FRG

Froning, M., Dipl.-Ing.
Forschungszentrum Jülich GmbH, Institut für Angewandte Physikalische
Chemie, Postfach 1913, 5170 Jülich, FRG

Grimmer, G., Prof. Dr.
Biochemisches Institut für Umweltcarcinogene, Lurup 4, 2070 Großhansdorf,
FRG

Herber, R.F.M., Dr.
Coronel Laboratorium Faculteit, Faculteit Geneeskunde, Academisch Medisch
Centrum, Meibergdreef 15, 1105 AZ Amsterdam, The Netherlands

Ihnat, M., Dr.
Land Resource Research Centre, Agriculture Canada, Ottawa, Ontario,
K1A OC6, Canada

Irgolic, K.J., Prof. Dr.
Institut für Analytische Chemie, Karl-Franzens-Universität Graz, 8010 Graz,
Austria

Jacob, J., Prof. Dr.
Biochemisches Institut für Umweltcarcinogene, Lurup 4, 2070 Großhansdorf,
FRG

Kemper, F.H., Prof. Dr.
Institut für Pharmakologie und Toxikologie, Westfälische
Wilhelms-Universität, Domagkstr. 12, 4400 Münster, FRG

Kettrup, A., Prof. Dr.
Forschungszentrum für Umwelt und Gesundheit GmbH (GSF), Institut für
Ökologische Chemie, Ingolstädter Landstr. 1, 8042 Neuherberg, FRG

B.J. Koster
Chemical Science and Technology Laboratory, National Institute of Standards
and Technology, Gaithersburg MD 20899, USA.

May, K., Dipl.-Ing.
Forschungszentrum Jülich GmbH, Institut für Angewandte Physikalische
Chemie, Postfach 1913, 5170 Jülich, FRG

Müller, C., Dr.
Institut für Pharmakologie und Toxikologie der Universität Münster,
Umweltprobenbank, Domagkstraße 12, 4400 Münster/Westfalen, FRG

Okamoto, K., Prof. Dr.
Pharmac. Analyt. Chemistry Laboratories, Faculty of Pharmaceutical Sciences,
The University of Tokushima, Shou-machi 1-78, Tokushima, 770, Japan

Ostapczuk, P., Dr.
Forschungszentrum Jülich GmbH, Institut für Angewandte Physikalische
Chemie, Postfach 1913, 5170 Jülich, FRG

Oxynos, K., Dr.
Forschungszentrum für Umwelt und Gesundheit GmbH (GSF), Institut für
Ökologische Chemie, Ingolstädter Landstr. 1, 8042 Neuherberg, FRG

Padberg, S., Dr.
Forschungszentrum Jülich GmbH, Institut für Angewandte Physikalische
Chemie, Postfach 1913, 5170 Jülich, FRG

Reisinger, K., Dr.
Institut Dr. Luss, Schönbornstr. 34, 8730 Bad Kissingen

Rossbach, M., Dr.
Forschungszentrum Jülich GmbH, Institut für Angewandte Physikalische
Chemie, Postfach 1913, 5170 Jülich, FRG

Schladot, J.D., Dr.
Forschungszentrum Jülich GmbH, Institut für Angewandte Physikalische
Chemie, Postfach 1913, 5170 Jülich, FRG

Schmitzer, J., Dr.
Forschungszentrum für Umwelt und Gesundheit GmbH (GSF), Institut für
Ökologische Chemie, Ingolstädter Landstr. 1, 8042 Neuherberg, FRG

Schwuger, M.J., Prof. Dr.
Forschungszentrum Jülich GmbH, Institut für Angewandte Physikalische
Chemie, Postfach 1913, 5170 Jülich, FRG

Stone, S.F., Dr.
Hahn-Meitner-Institut Berlin GmbH, Bereich Strukturforschung,
Abt. 4: Spurenelemente in Gesundheitund Ernährung, Glienickerstr. 100,
D-1000 Berlin 39.

Tavares, T.M., Prof. Dr.
Instituto de Quimica/UFBa, Campus Universitario da Federacao, s/n, 40210
Salvador, Bahia, Brazil

S.A. Wise
Center for Analytical Chemistry, National Institute of Standards and
Technology, Gaithersburg MD 20899, USA.

Zajc, A., Dr.
Institut für Wasserchemie, TU München, Marchioninistr. 17, 8000 München 70,
FRG

R. Zeisler
International Atomic Energy Agency, Agency's Laboratory Seibersdorf,
P.O. Box 100, A-1400 Vienna, Austria.

Contents

5. Organic Analytical Approaches

6. Inorganic Analytical Approaches

7. Future Developments

1 Introduction

1.1 Preface

M.J. Schwuger

The Environmental Specimen Bank is a repository of representative environmental specimens for safe long-term storage over decades and centuries without any chemical change in the constituents. It represents the modern form of a systematically designed collection which will permit comparative analyses and evaluations of chemicals in the future. The aims are:

- the determination of selected chemical compounds at the time of storage,
- comparative investigations with new methods for chemicals which, at the time of storage, could not be determined or were not recognized as important,
- observation of trends in the environment using authentic material from the past and
- documentation of long-term changes.

Environmental specimen banking is thus suitable for identifying environmental changes and initiating necessary measures of remediation. It may be used to identify problems, study correlations between cause and effect and determine the effectiveness of legislative measures as well as to recommend the activities required.

This is not only ecologically important, but also relevant for man, since he is the last member in the food chain and is therefore affected by all compartments of the environment. For this reason, two banking systems were established in Germany, one for environmental specimens (Jülich), and the other for human specimens (Münster).

Since the concentration of pollutants in the individual compartments of the environment (e.g. water, soil, air) or single substrates does not provide any information about their accumulation in the food chain, it is also necessary to analyse representative biological specimens, ranging from microorganisms up to man. Figure 1 schematically shows the enrichment and depletion chains for which representative specimens from all spheres are collected, processed, analysed and stored in the Environmental Specimen Bank.

Figure 2 schematically shows, using mercury as an example how heavy metals can accumulate in the food chain.

Since the concentration of environmentally relevant substances is extremely low in the substrates, significant results can only be obtained by means of

Specimen Banking
Rossbach/Schladot/Ostapczuk (Eds.)
© Springer-Verlag Berlin Heidelberg 1992

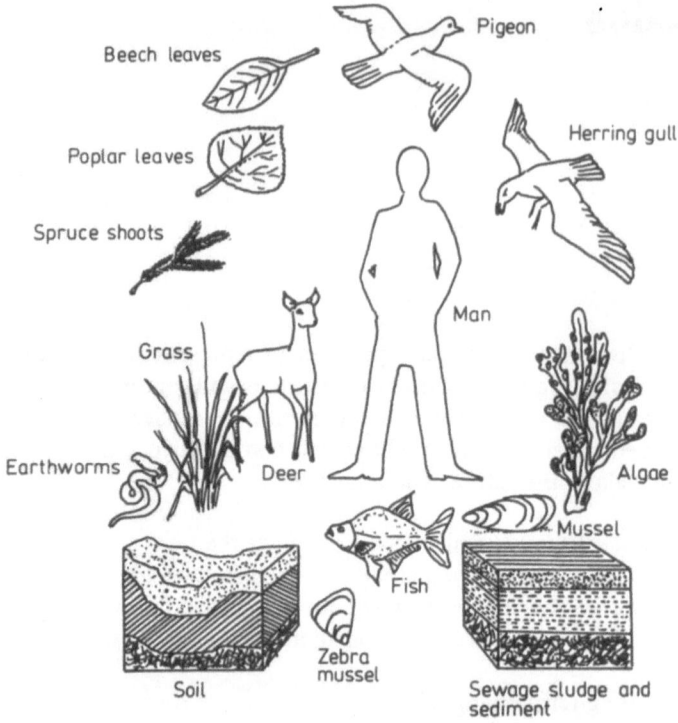

Fig. 1. Schematic view of different matrices collected for the Environmental Specimen Bank from various compartments of the ecosphere

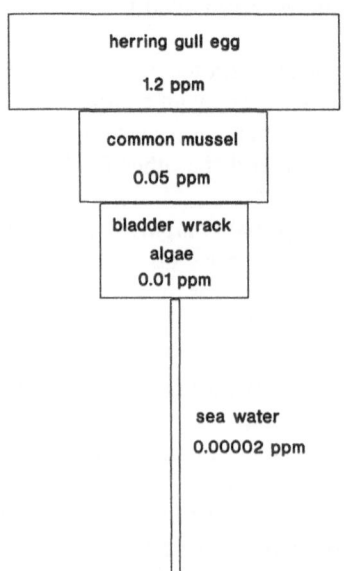

Fig. 2. Accumulation pathway of mercury in marine compartments

specifically designed sampling procedures and the most up-to-date methods of analysis.

It was recognized at an early stage that specimens from natural history collections do not permit reliable conclusions to be drawn, since neither sampling nor preservation and storage conditions ensure the significance of retrospective investigations.

Special criteria are necessary for the selection of sampling areas apart from the selective choice of indicators for the analysis of environmental concentrations. Important factors are the ecological representativeness, the size of the area, the stability of soil use, accessibility to the sampling areas and the legal conditions for long-term land use.

Trace analysis in the environment requires particularly precise – primarily physico-chemical – methods of investigation, which have only been developed in recent times for highly sophisticated environmental analyses and are constantly being improved and extended. In addition to the analytical characterization of organic and inorganic substances in specimens from the environment, it is also indispensable to characterize their biological and physical properties. Every matrix requires a special sampling and processing procedure which, as a rule, clearly differs from conventional methods of investigation for corresponding substances.

It was only possible to build up environmental specimen banking after modern chemistry had created these most diverse prerequisites.

It was then demonstrated that many measurements had not been sufficiently exact 20 and more years ago in order to document significant changes. This led to the decision to establish Environmental Specimen Banks (ESB) in the Federal Republic of Germany and the USA. It is currently being planned to build up similar systems in other countries in view of the success and interesting results achieved to date in the existing banks, especially taking the German Environmental Specimen Bank as a model.

The development of Environmental Specimen Banking in the Federal Republic of Germany is closely linked with the name of Markus Stoeppler. Since the early seventies, he has made pioneering contributions to sampling, processing, storage and especially to analysis techniques for various types of specimens. He belongs to the founders of the Environmental Specimen Bank in Jülich where all selected types of specimen from all defined sampling areas are currently being stored.

The particular advantage of Environmental Specimen Banking at the Research Centre Jülich comprises the versatile analytical possibilities and far-reaching environmental activities for studying both environmental pollution and the structure-effect relationships in the environment. Optimum results have been achieved within a short time due to the fact that these activities are embedded in a major environmental programme.

After more than 10 years of Environmental Specimen Banking in the Federal Republic of Germany and after the retirement of its mentor, Markus Stoeppler, it seems appropriate to present a survey of selected results and plans of Environmental Specimen Banking to the public.

1.2 Editorial Foreword

M. Rossbach, J.D. Schladot, P. Ostapczuk

Planning, collecting and editing the contributions of this book dedicated to Markus Stoeppler's great merits and his role in the development and organization of the specimen bank idea in general and the Jülich activities in particular was a great pleasure and a new experience for all of us. New perspectives are emerging from such a broad scale approach: Specimen banking (in highly industrialized countries) has clearly left infancy behind and has become a mature scientific enterprise. If one takes the establishment of more precise "reference ranges" in human fluids and tissues, the real-time and trend monitoring using biological indicator organisms, development of more accurate and reliable analysis systems for organic and inorganic constituents of the stored samples, or the provision and application of reference samples etc. specimen banking has shown to be a particularly reliable instrument for ecological, toxicological and analytical research. It is not that the idea of specimen banking has still to be justified as numerous examples exist now where banked samples helped to evaluate trends and risk assessment (see e.g. R.E. Lee, 1990, in: R. Zeisler, V.P. Guinn, Nucl. Anal. Meth. in the Life Sciences, Humana Press, Clifton NY or several contributions in this book) but extensive use of the valuable sample archive and the emerging data has so far not been made by the legislative and thus stable funding commitment – an indispensable prerequisite for such a long term project – has not yet been established.

The associated scientists, however, greatly appreciate the banking facilities and take every effort to spread the idea. Implementation of the ESB concept to developing countries is a hot topic (as will be discussed at the first Int. Symp. on Biological and Environmental Specimen Banking to be held from 22 to 25 Sept. 1991 in Vienna, Austria) as well as the determination of the chemical species of toxic elements present in the biological samples. ESB material is the most appropriate for such investigation because it is unaltered, freshly homogenized, and deep frozen. Labile organometalic compounds and metalo-enzymes will most probably be not destroyed. Another future perspective might be (and has already been demonstrated in the case of the Chernobyl accident) the retrospective analysis of single isotopes where the fund of banked samples could turn out to be of unexpected value.

Nevertheless many open questions concerning specimen types, collecting areas, storage conditions, and organisation of the data pool need to be

Specimen Banking
Rossbach/Schladot/Ostapczuk (Eds.)
© Springer-Verlag Berlin Heidelberg 1992

addressed further. Direct sampling of air specimens would be desirable particularly for atmospheric and climate research purposes but no practical storage ability is known. Atmospheric particulates can be collected in small quantities as individual samples but the banking concept foresees large amounts of homogenized materials. Therefore this matrix is not included in pure state but as precipitate on bioindicators like spruce shoots or poplar leaves.

Specimen banking activities in Germany are now being more strongly directed towards the eastern part of the country (former GDR). Thus the ideas for environmental protection underlying the ESB concept will hopefully spread further to the east as enviornmental impact caused by human activities is by no means a national but rather a more or less global affair. Consequently international bodies such as the European Community or UN organisations should finally be asked to take over their responsibilities for the improved environmental quality of their member states.

1.3 Markus Stoeppler – Impressions of his Life with Science

G. Erdtmann

Markus Stoeppler was born on May 22nd 1927 in a Swabian parsonage at Stuttgart. He spent his childhood in several places in the southwest of Germany – in the Palatinate, in the Saarland, and in Wurttemberg – according to the parishes where his father had been transferred. Before he could finish the secondary school, it was in 1944, the last year of World War II, the pupil was conscripted as "Flakhelfer" into the German army. Therefore, he passed the high school examination in Calw (Wurttemberg) only in 1947. Obviously, at that time he had already aspired to become a chemist, but after the war it was nearly impossible to get a place at one of the German universities. Some of them were totally destroyed, many were bombed, and thousands of soldiers coming back from war and captivity also wanted to attend university. Thus, Markus Stoeppler entered the Badische Anilin- und Sodafabrik, now BASF, in Ludwigs-hafen/Rhein and became a laboratory worker in order to pursue a career in chemistry.

In May 1946, the University of Mainz, which had been closed by the French army of occupation under Bonaparte in 1798, was reopened by the commander of the French occupying forces – General Koenig. In 1949, Markus Stoeppler matriculated here as a student of natural science. After an interruption, caused by the death of his father, he finished his studies in 1958. The practical work required for the diploma degree was carried out at the Institute of Inorganic Chemistry. The main fields of research in this institute at that time were *radio- and nuclear chemistry* under Prof. F. Straßmann and *trace and microanalysis* under Prof. W. Geilmann. These areas of research and their applications were decisive for the further scientific career of Markus Stoeppler.

The title of Markus Stoeppler's diploma thesis was "Search for Unknown Alpha-Emitters" a subject on which Prof. Straßmann had initiated several diploma works. On the chart of the nuclides no white spots could be seen except at the edges, where long-lived isotopes were not to be expected. But Prof. Straßmann – after the unexpected detection of nuclear fission – did not trust seemingly well proved theories and traditional opinions. Accordingly, he suspected the existence of further natural radioactive nuclides and this opinion was supported by the several papers describing unexplainable α-rays of low energies emitted by different geological materials such as biotite and cordierite, and ores, products and by-products of zinc and copper metallurgy. Following the bio-

Specimen Banking
Rossbach/Schladot/Ostapczuk (Eds.)
© Springer-Verlag Berlin Heidelberg 1992

Dr. M. Stoeppler

graphy of F. Krafft [1], he even wanted not to exclude, that possibly all natural elements were radioactive with extremely long half-lifes. Markus Stoeppler finished his diploma thesis after the prescribed time without having detected any unknown α-emitting nuclide [2]. However, he had not only learned delicate nuclear detection techniques – nuclear track emulsions and α-ionization chambers were mainly used during the work – but he also became familiar with techniques of chemical trace analysis and its problems such as cross contamination and memory effects. This experience was of great value for his later work.

Markus Stoeppler's doctoral thesis was also concerned with natural radionuclides. By this time these were well known ones: ^{7}Be, ^{35}S, and ^{210}Po. Their appearance, the concentrations and the seasonal variations in the atmosphere had to be measured using air filter and precipitate water samples, and appropriate methods for their separation and measurement had to be developed. This was the experimental part of the doctoral thesis, which was carried out at the University of Heidelberg in the II. Physical Institute under Prof. Haxel and

---

[1] F. Krafft *Im Schatten der Sensation – Leben und Wirken von Fritz Straßmann* Verlag Chemie Weinheim 1981, p. 162.
[2] Since 1958, indeed, four natural isotopes have found to be α-radioactive. The predecessors of Nd-144, namely Sm-148 and Gd-152, Hf-174, and Os-186, the daughter of Pt-190.

G. Schumann. Markus Stoeppler received his title Dr. rer. nat. under Prof. Straßmann at the Johannes-Gutenberg-University at Mainz in 1963.

In the same year he joined the Isotopen-Studiengesellschaft in Karlsruhe. Here he investigated the composition of various phosphorus–vanadium–molybdenum–complexes which were to find analytical use in a substoichiometric determination procedure of phosphorus in steel, and the radionuclides ^{32}P and ^{99}Mo were used as radioactive tracers. In another study he was engaged in the development of an energy dispersive X-ray fluorescence spectrometer with a radionuclide source – the sources consisted of tritium loaded disks of titanium or zirconium – which was to be used for the on-line determination of light elements in streaming gases in industrial processes.

After a short period at Unilever in Hamburg Markus Stoeppler joined the Nuclear Research Center (KFA) Jülich in 1966. Meanwhile he had married and was the father of three daughters. The KFA Jülich had been founded in 1956 and was still in a state of expansion. Its aim at that time was the development of nuclear techniques, mainly for power reactors but also for applications within the fields of natural and medical science. Here was a wide field for activities in radiochemical analysis. Initially, he entered the Institute of Reactor Components and in 1969 he changed over to the Central Laboratory of Chemical Analyses, the head of which was Dr. H.W. Nürnberg.

One of the large projects of the KFA Jülich was the development of the Thorium High Temperature Reactor (THTR). Markus Stoeppler started to develop analytical procedures for irradiated fuel elements of the pebble-bed reactor. The ball-shaped fuel elements contained the fuel in the form of "coated particles". Procedures to dismantle and dissolve these particles and to separate uranium from the fission and activation products were required. The quantitative determination of uranium, leachable from the pyrocarbon coating by nitric acid ("leach test"), or a mass spectrometric measurement of the abundance of its isotopes ("burn-up determination") were the final steps of these procedures.

An outstanding property of the fuel elements of the high temperature reactor is their high impermeability toward the fission products produced during operation and this was to be confirmed by numerous tests. Markus Stoeppler developed procedures to separate and determine ^{89}Sr and ^{90}Sr, which were of special radiotoxicological importance and which could not be measured simply by γ-ray spectrometry. The chemical treatment of the strongly radioactive samples had to be carried out in lead shielded cells and special, often remotely controlled apparatuses were needed. We can assume, that Markus Stoeppler, who during his free time liked to build small models of aircrafts, especially liked and enjoyed the construction of this neat and sophisticated equipment.

In 1971, the Central Laboratory of Chemical Analysis was reorganised under the name Central Institute of Analytical Chemistry and the directorate of Prof. H.W. Nürnberg. Markus Stoeppler became head of a new Division called "Tracer and Separation Methods". Here, the main field of research was micro and trace element analysis and, among other things, separation procedures were developed for the determination of palladium in platinum and of uranium by neutron activation analysis. But now he had directed his attention more and more to nonradioactive analytical methods. With improvements in the automatic sample addition procedure for graphite cuvettes he introduced himself into the community of the atomic absorption spectrometrists.

The plans for the future, however, were much more extensive. The detrimental impact of human activities – industry, traffic, intensive agriculture – on the natural resources of life had become so obvious (Rachel Carson's book "The Silent Spring" had been a bestseller during the sixties in Germany), that many scientists felt alarmed and were urged to take action. Hence, a great share of the research capacity of the institute was shifted into the new environmental activities and Markus Stoeppler's [3] group took the part of developing analytical procedures for inorganic biocides. Lead, cadmium and mercury, elements the toxicity of which was well known, were among the first candidates to be investigated. It was necessary to study their behaviour in the environment, their pathways from the atmosphere to the soil and to surface and underground water, from there into plants, animals, and food, and finally into man. The transfer factors as well as the toxicity of the harmful elements could not be simply related to the elemental concentrations. It was found, that these properties strongly depend on the chemical form in which the heavy metals are bound. The identification and quantification of these compounds, the techniques of "speciation", required additional analytical methods. Other procedures had to be introduced to improve the quality of the results depending on the types of samples and the elements to be determined. Thus, the name Markus Stoeppler increasingly appears on papers dealing with polarography, voltammetry, isotope dilution mass spectrometry, emission spectrometry, and activation analysis, and the applications of these methods in environmental sciences.

During these studies it was realized that toxic heavy metals were found nearly everywhere and often it could not be determined whether their occurence was the consequence of human activities, i.e. from industrial emissions, or if they were deposited by natural processes. This was the situation because there were no analytical data of corresponding sites and specimens of the period before the industrial era. And if such data were available, then they were obtained by analytical procedures much less sensitive than those presently in use, so that the results were not comparable. If the development of pollution was to be followed at least in the future and if the effect of preventive measures, meanwhile introduced and prescribed by law, were to be monitored, then the environmental scientists required a data bank of reliable analytical results. However, this was a difficult task. Comparisons of results obtained in different laboratories with different techniques often did not agree at all. Although in most of the cases such discrepancies could be explained and overcome after additional research work, there remained the experience that analytical methods and analytical scientists

---

[3] Since this book is edited in honour of Markus Stoeppler, his name appears sometimes where, indeed, a number of co-workers, contributors and predecessors from in and outside the Institute of Applied Physical chemistry of the KFA Jülich should have been mentioned.

were not à priori infallible. Furthermore, new pollutants were expected to occur and new analytical procedures had to be invented to measure these as well as the old ones, and these new data, of course, could not be included in a data set from the present. Therefore, besides the analytical data, the specimens themselves had to be archived to freeze the present state of pollution at least for some selected sites and types of environmental specimens. This had to be done in the true sense of the word. Because most specimens were taken from living matter they were subject to spoiling.

Thus the idea of an Environmental Specimen Bank was born, where samples are stored at very low temperatures so that the speed of chemical reactions and biological processes was reduced to practical zero. The samples should be well analyzed and characterized by the techniques presently available. They should serve the coming generations as original sources like the incunabula in the libraries of former centuries. This idea was realized with the impetus of Prof. Nürnberg and in co-operation with many other colleagues of the institute, from which mainly Dr. Dürbeck, whose group dealt with the hazardous organic chemicals, should be mentioned. A pilot project was initiated under the leadership of the Umweltbundesamt and in 1980 the construction of a building for this Environmental Specimen Bank was started at the KFA. This project attracted worldwide attention and the scientists thus engaged, like Markus Stoeppler, became requested speakers at international conferences on environmental chemistry. Guests from many countries came to the institute to work with the teams engaged in this project. New ideas were proposed and it became more and more difficult to keep an overview of all the new topics treated in the institute.

Looking through the papers of Markus Stoeppler, the following points appear to describe the activity during this period. The first one is the continuous work on the improvement of the analytical methods with respect to practicability and to accuracy. The screening of polluted and unpolluted areas required the analysis of large numbers of samples and therefore the methods, especially the sample preparation procedures, had to be automatized to allow maximum sample throughput. Analyses had also to be carried out even under hard conditions in a field laboratory and hence the procedures had to be made simple and robust. Improvement of accuracy, i.e. with trace element analysis, mainly the exclusion of all possible sources of contamination, which had made many analytical results from the pioneer time of environmental analysis obsolete. One of the means of quality assessment in analytical chemistry is the consequent use of reference materials, and therefore Markus Stoeppler engaged himself in the development of reference materials appropriate for trace element determinations in biological samples.

The second point involves cadmium. This rare element was soon found to be widely distributed in the environment, although its industrial use and importance was believed to be very limited. It could be sensitively determined by polarography. Markus Stoeppler contributed to the improvement of the determination by atomic absorption spectrometry, so that these findings could be confirmed by independent methods, which were also used to establish a data base for investigations on the ecotoxicological effects. Markus Stoeppler gained reputation as an expert in the trace analysis of cadmium and was invited to write review articles on this topic.

The third point concerns mercury. The metal and its inorganic, ionic compounds are much less toxic than the organic compounds. Especially methyl mercury compounds turned out to be very toxic and they were found not only in areas polluted by industrial emissions but also in "clean air areas". The question of how these compounds are built up and decomposed by natural, biochemical processes has strongly engaged him; he and his co-workers have contributed substantially to the research in this area which, however, still requires a lot of work in order to understand all the reactions and their kinetics.

The last point to be mentioned is bioindicators. Since the space in an Environmental Specimen Bank is very limited, the samples to be stored have to be carefully selected. It seems comparatively simple (although it is complicated enough) to take samples of soil and water to archive the present state of these compartments of the environment, and to collect leaves from trees and mosses, or earthworms, to indicate local immissions, however, it seems, impossible to store useful samples of air. Thus, the idea was developed, and here as a co-worker Edmund Hahn is mentioned, to use feathers of birds as bio-indicators. Their plumage is in continuous and intimate contact with the air and filters out many of the fine, suspended particles, which are deposited on the feathers. These can be collected without sacrificing the birds. Some birds have very small habitats, so that they can serve as very local indicators, whereas others have large territories and are useful for screening larger areas. Migrating birds, spend their lives in very different parts of the world and hence are useful for the global monitoring of pollutants.

Thus, Markus Stoeppler, who will retire from the KFA after his 65th birthday, has become a highly reputed scientist in the analytical chemistry of environmental inorganics. He leaves a well developed field and a capable and motivated team, so that we can expect that his work will be successfully continued at the Institute of Applied Physical Chemistry.

2 Specimen Banking

2.1 Specimens of Human Origin for Biomonitoring

F.H. Kemper

The pattern of human matrices useful for biomonitoring and therefore stored in the Environmental Specimen Bank (ESB) includes specimens from autopsy material (liver and adipose tissue) as well as 'available organs' from living individuals: whole blood, blood plasma, and urine are used for inorganic and organic sample characterization: The compounds and analytical methods are listed. Advantages and disadvantages of the easy available specimen hair are discussed. Evaluation and judgement of analytical results must be based on reference ranges. ESB data from healthy probands are given for accidental and essential trace elements in whole blood, urine and scalp hair in comparison with collected literature results. Specific toxicokinetic indicator specimens are listed for some trace elements of environmental importance; toxicodynamic indicators are added.

Using the mentioned scientific tools human biomonitoring may be applicated to ensure acute environmental exposure (e.g. to industrial emissions) as well as for trend analysis: As a result of 7 years biomonitoring of pentachlorophenol (PCP) or lead content in body fluids a distinct drop of the concentrations of these xenobiotics could be observed due to legislative restriction of PCP and Pb use.

Introduction

Humans are a target of numerous environmental and also anthropogenic chemical influences (Fig. 1) [1]. Humans are in contact with all parts of the biosphere and are one of the final links of the food chain. Consequently specimens of human origin were integrated into the logistic scheme of the Environmental Specimen Bank (ESB) at the beginning of the institution [2].

Among the objectives of banking and characterization, human specimens can be of use to evaluate the xenobiotic background level, can detect burden trends, and have indicator functions in available matrices for conclusions on whole body exposures.

Available Specimens from Living Individuals

Human specimens are of great importance for short time ("Real time") monitoring (RTM) as well as for longtime trend analysis in the fields of health protection. Thus it is necessary to collect and store both, (i) RTM samples from living people for measuring the actual contents (burden) in available organs in the population and (ii) another number of organs gained from autopsies. By the

Specimen Banking
Rossbach/Schladot/Ostapczuk (Eds.)
© Springer-Verlag Berlin Heidelberg 1992

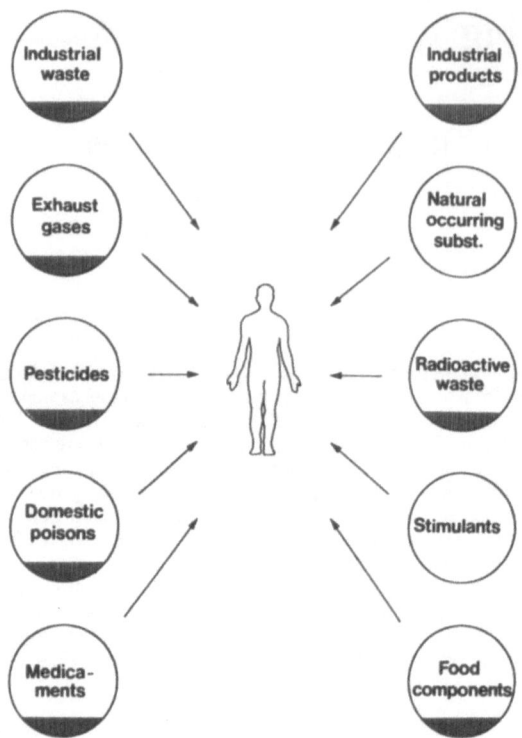

Fig. 1. Chemical environmental influences on man (from [1])

end of 1990 more than 280000 samples from healthy people had been collected under these premises in the ESB Münster (Environmental Specimen Bank for Human Tissues, Münster/FRG). In RTM twice a year each collective consists of 100–150 volunteers, who have to answer a detailed questionnaire to get sufficient information about their individual living area, smoking, drinking, food, drug habits, and other personal data, which may influence xenobiotic kinetics.

From these comparable groups the following specimens are collected:

- Whole blood (80 ml in 5 subsamples, anticoagulated by heparine).
- Blood plasma (15 ml in 3 subsamples).
- 24 h collected urine (divided in 16 subsamples).
- Spontaneous saliva (2 ml in 2 subsamples).
- Scalp hair (0.5 g cut in 2.5-cm-segments).
- Pubic hair (0.5 g in 2 subsamples).

Sampling is done following specific Standard Operation Procedures (SOP) in order to minimize external contamination by ubiquitous trace substances and to ensure the reproducibility of sample collection and the comparability of the respective analytical results.

Table 1. Advantages and disadvantages of hair as biomonitoring species

Advantages	Disadvantages
– Sampling without problems	Trace element content may be influenced by:
– Storage at room temperature possible	– age, sex, hair colour, ethnic origin
– Trace element content higher than in blood or urine	– different growth phase of single hairs
– Matrix less complex than blood or urine	– localization on head
	– preparation for analysis (washing procedure)
– Longitudinal analysis gives a 'trace element history'	– exogeneous contamination

Among the above listed specimens blood, blood plasma or serum and urine are well established samples for diagnosis in clinical and environmental medicine. Hair is easily available, but interpretation of analytical data must consider known influence factors on the content of trace compounds (Table 1) [3]. Consequently hair samples should be used only in connection with other specimens for biomonitoring purposes. The role of saliva in biomonitoring is not yet well established. Currently the evaluation of a sufficient data pool is done with high priority in our own laboratory.

In special groups of persons other suitable types of matrices are also available. Human milk is a convenient specimen for biomonitoring of inorganic compounds. Because of its high lipid content it is also used for organic xenobiotics; a 20 ml sample is sufficient for a routine monitoring run. Moreover time-dependent observations can be done during the whole lactation period. Seminal fluid may be used for exposure tests in the inorganic field. Data are published, which indicate a drop in lead content in seminal fluid in 1989 compared to 1987 [4]. This may indicate a decreased environmental Pb exposure. Feces, nails or cerebrospinal fluid may be useful as diagnostic tools only in acute or chronic intoxications (e.g. with thallium or arsenic). In routine biomonitoring these samples have not proven advantageous and thus are of marginal interest.

Autopsy Material

During the initial run of the ESB pilot phase up to 31 different tissue types from autopsy material were characterized and stored, according to age, sex and autopsy diagnosis. Finally two specimens were selected for continuous biomonitoring and storage:

a) liver tissue (main metabolizing organ/high lipid content/increased xenobiotic content/suitable for analysis of organic and inorganic compounds);
b) adipose tissue (tissue with highest lipid content/enrichment of persistent organic compounds).

From each of these two matrices 250 g homogenized tissue per donor (one liver lobe/subcutaneous adipose tissue of the abdomen region) are stored and divided into 15–20 subsamples. The samples are normally taken from humans who died of accident or acute coronary heart disease, thus avoiding unknown influences of chronic diseases on distribution or metabolism of xenobiotic compounds. For special questions e.g. lifelong body burden specimens obtained from elderly people are also examined and stored. According to recent analytical methodology and knowledge a single subsample is sufficient for the determination of all known trace compounds including 'subtrace' substances (e.g. dibenzodioxines and dibenzofuranes in adipose tissue).

Some of the persistent xenobiotics exhibit an exposure/time gradient with distinct increased levels in older age. Figure 2 shows the age dependence of hexachlorobenzene (HCB) and total dichlorodiphenyltrichloroethane (DDT) contents in human adipose tissue [5]; in diminished form this can be seen in liver tissue too. In Fig. 3 the age dependence of cadmium in liver and kidneys of children is presented [1]. A rapid load from the first days of life is obvious. Beyond the age of 20 further increase of Cd content is seen: up to 45 μg Cd/g in the kidney (at age 60) and 2.0 μg Cd/g in the liver (at age 80).

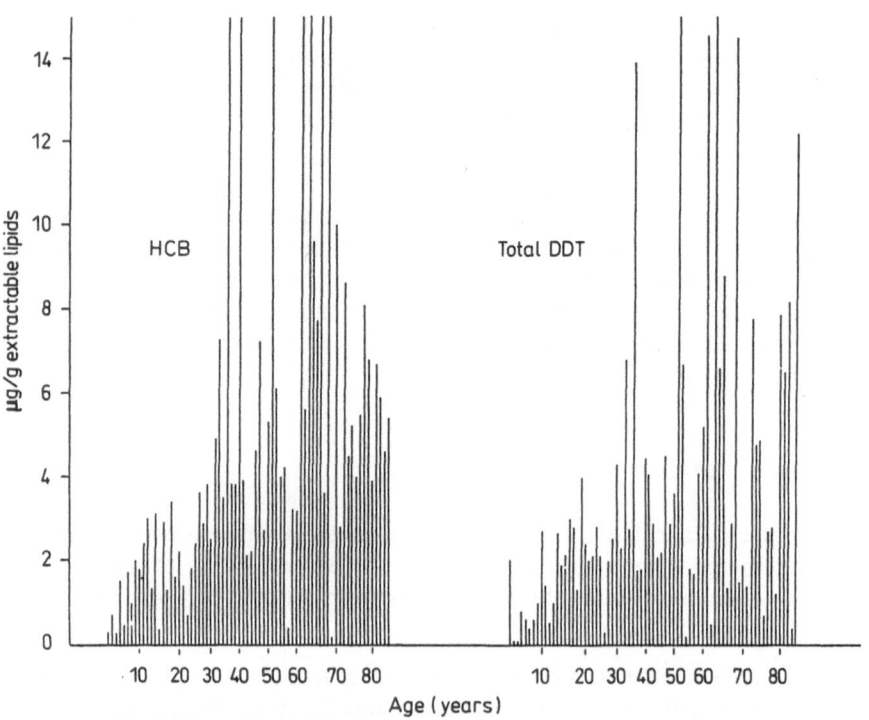

Fig. 2. Age dependence of hexachlorobenzene (HCB) and total dichlorodiphenyltrichloroethane (DDT) content in human adipose tissue (from [5])

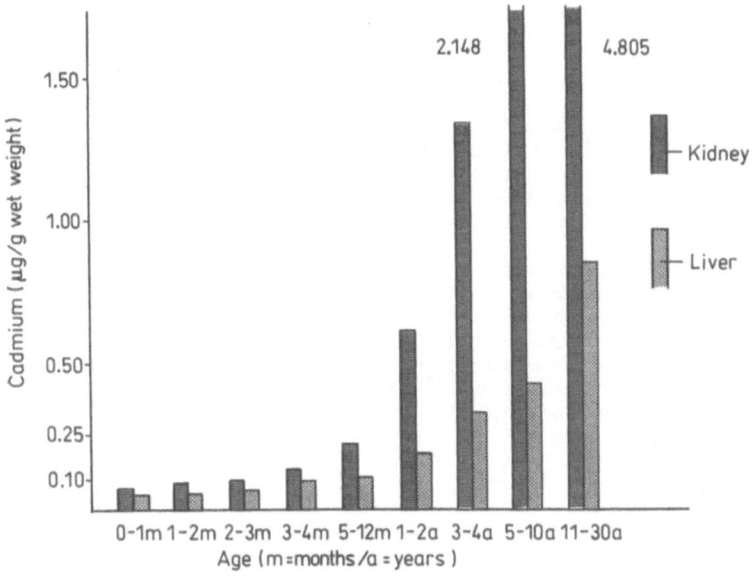

Fig. 3. Age dependence of cadmium content in liver and kidneys of children (from [1])

Analytical Verification of Biomonitoring

Apart from the full anamnestic history the characterization of the Environmental Specimen Bank samples must include as many analytical data as possible. Thus multicomponent methods are integrated in the analytical run. In the organic field capillary gas chromatography with different columns and detection conditions is the method of choice. If necessary and especially in very low concentration ranges, compound identification is done by gas chromatography/mass spectrometry combination, using e.g. Reconstructed-Single-Ion-Detection.

Inductively coupled plasma emission spectroscopy (ICP) is one of the most useful inorganic multielement methods, yielding a lot of information about the sample within a short time. Unfortunately the content of most accidental trace elements in human samples are below the detection limits of ICP. Enrichment of trace metals before the analytical run involves the risk of contamination but the recently developed minimum handling systems (flow injection) may expand the possibilities of this method.

Electrothermal, hydride generation and cold vapor atomic absorption spectrometry (ET-, HY-, CV-AAS) complete the inorganic characterization of the samples [6]. Table 2 lists the compounds analyzed for characterization purposes of human specimens in the ESB program. Figures 4 and 5 point out the standardized analytical run schemes for whole blood/blood plasma and human milk.

Table 2. Scope of biomonitoring parameters for characterization of human samples

a) Organochlorine xenobiotics

2,2-Bis-(4-chlorophenyl)-1,1,1-trichloroethane	(p,p'-DDT)
2-(2-chlorophenyl)-2-(4-chlorophenyl)-1,1,1-trichloroethane	(o,p'-DDT)
2,2-Bis-(4-chlorophenyl)-1,1-dichloroethane	(p,p'-DDD)
2,2-Bis-(4-chlorophenyl)-1,1-dichloroethene	(p,p'-DDE)
2-(2-chlorophenyl)-2-(4-chlorophenyl)-1,1-dichloroethene	(o,p'-DDE)
Hexachlorobenzene	(HCB)
alpha-1,2,3,4,5,6-Hexachlorocyclohexane	(α-HCH)
beta-1,2,3,4,5,6-Hexachlorocyclohexane	(β-HCH)
gamma-1,2,3,4,5,6-Hexachlorocyclohexane	(γ-HCH)
1,4,5,6,7,8,8-Heptachloro-2,3,3a,4,7,7a-hexahydro-2,3-epoxi-4,7-methanoindene (Heptachloroepoxid)	(HE)
1,2,3,4,10,10-Hexachloro-6,7-epoxi-1,4,4a,5,6,7,8,8a-octahydro-endo-1,4-exo-5,8-dimethanonaphthaline (Dieldrin)	(HEOD)
Pentachlorophenol	(PCP)
2,4,4'-Trichlorobiphenyl	(PCB 28)
2,2',5,5'-Tetrachlorobiphenyl	(PCB 52)
2,2',4,5,5'-Pentachlorobiphenyl	(PCB 101)
2,2',3,4,4',5'-Hexachlorobiphenyl	(PCB 138)
2,2',4,4',5,5'-Hexachlorobiphenyl	(PCB 153)
2,2',3,4,4',5,5'-Heptachlorobiphenyl	(PCB 180)

b) Accidental trace elements

Lead	Arsenic	Thallium	Cadmium
Mercury	Silver	Antimony	Tin
Aluminum	Strontium	Barium	Beryllium
Boron			

c) Essential trace elements

Copper	Zinc	Iron	Manganese
Chromium	Selenium	Nickel	Vanadium
Fluorine			

d) Bulk elements

Calcium	Magnesium	Sodium	Potassium
Phosphorus	Sulphur		

e) Physiological organic compounds

Total protein	Creatinine	Glucose	Uric acid
Cholesterine	Triglycerids	LDH	SGOT
SGPT	γ-GT	Phosphatase	

Reference Ranges for Trace Compounds in Human Specimens

The evaluation and judgement of analytical data must be based on reference ranges. Particularly in the case of trend analysis with only slight or even marginal changes the knowledge of a quantil distribution with high precision is required. It is one of the objectives of the ESB Münster to collect as many

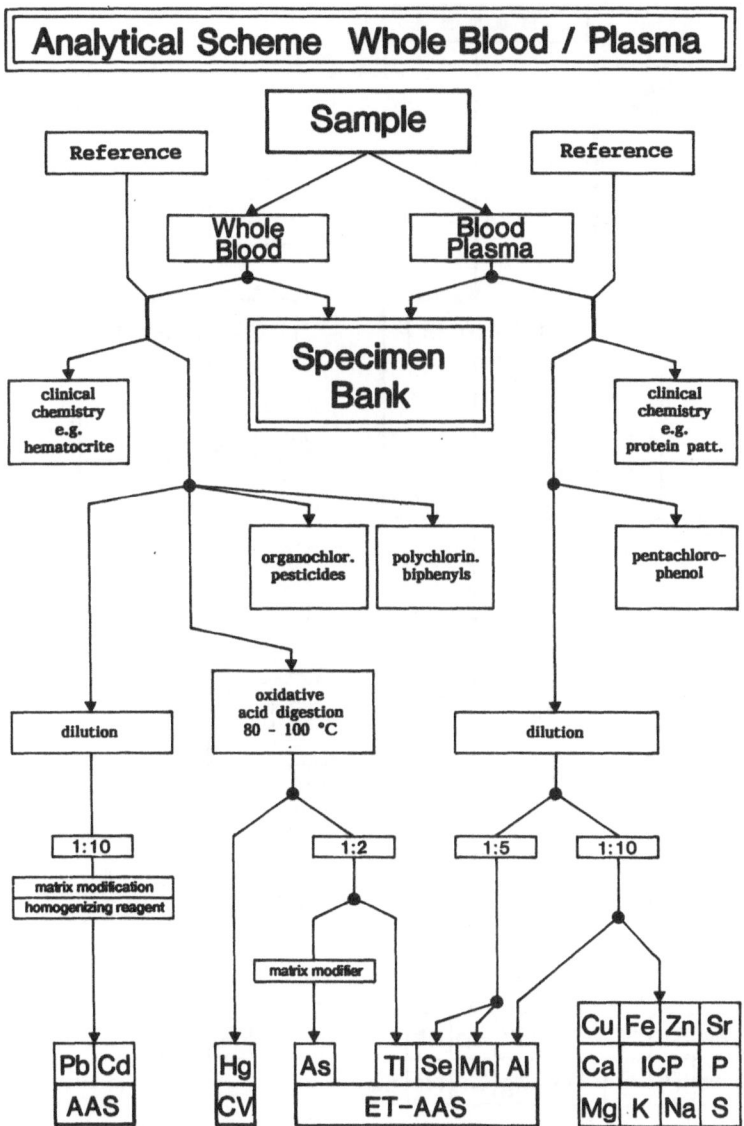

Fig. 4. Analytical run scheme for whole blood/blood plasma (from [2])

reliable data as possible from healthy probands to set up these reference values. Published 'normal ranges' data must be used critically, if they are not based on the knowledge of the numerous influences on trace contents in human organs (Table 3).

ESB results used as reference ranges are compared with the collected literature data in Tables 4–6 for whole blood, blood plasma, urine, and scalp hair.

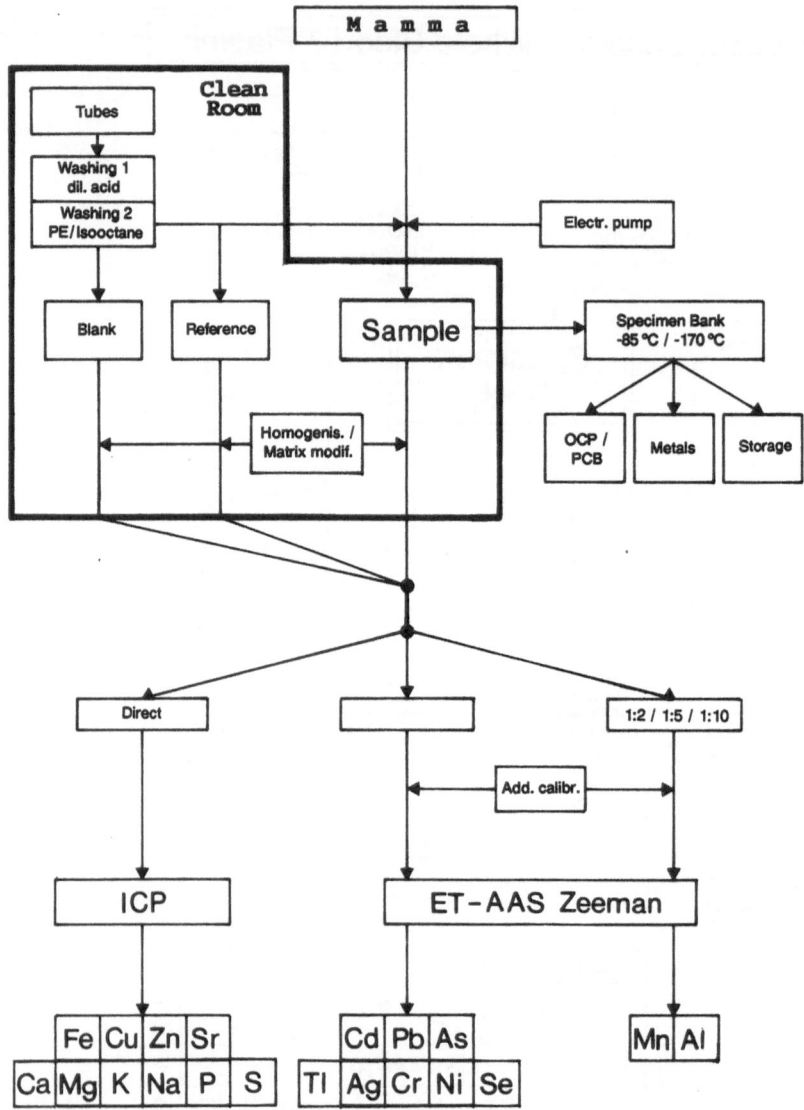

Fig. 5. Analytical run scheme for human milk

Indicator Function of Human Specimens

Biomonitoring data interpretation can only be carried out with a thorough knowledge of the importance and the indicator function of the analyzed matrix for the whole body burden, if 'available' specimens are used. In the field of organic xenobiotics, matrices with sufficient lipid content are useful. This is

Table 3. Influence on biological variability of trace components in human organs

Influencing factor	Cause
Endogeneous	
– age	changed kinetics/changes in tissue composition
– sex	differences in metabolism due to steroid hormones
– sampling time	circadiane rhythms
– race	metabolic differences
– diseases	changes in distribution
– position during sampling	e.g. different water content in blood plasma
Exogeneous	
– nutrition	different contents of trace compounds in food
– season/climatic zone	dependence of metabolic activity on climatic factors
– living area	industrial emission/agricultural and forestral use of xenobiotics/traffic
– profession	occupational exposure
– therapeutic treatment	changes in kinetics/chelating effects of drugs
Historic	
– analytical difficulties	evolution of physicochemical analytical methods

Table 4a. Reference ranges for trace elements in whole blood

Element	ESB results	Data from literature	Unit
Lead	(20) 25–120 (250)[a] Quant. 89% → 92.1 96% → 115.6 98% → 131.2	(10) 30–350 (565) Quant. 89% → 200 96% → 300 98% → 350	µg Pb/L
Cadmium	0.1–1.0 nonsmoker	(0.03) 0.3–1.2 nonsmoker 0.6–4.0 (7.5) smoker	µg Cd/L
Arsenic	0.5–2.0	(0.2) 1.0–10 (25)	µg As/L
Mercury	< 2.0 (3.0)	(0.1) 0.6–10	µg Hg/L
Thallium	< 2.0	≈ 0.5	µg Tl/L

[a] n = 630 (1986–1989).

Table 4b. Reference ranges for trace elements in blood plasma

Element	ESB results	Data from literature	Unit
Copper	(0.50) 0.80–1.30 (1.70)	(0.42) 0.70–1.50 (1.80)	mg Cu/L
Zinc	(0.50) 0.55–1.10 (1.30)	(0.50) 0.55–1.30 (1.75)	mg Zn/L
Selenium	50–120	(40) 60–120 (150)	µg Se/L
Iron	(0.45) 0.80–1.40 (2.0)	(0.40) 0.75–1.50 (1.70)	mg Fe/L

Table 5. Reference ranges for trace elements in urine

Element	ESB results	Data from literature	Unit
Zinc	(100) 150–600 (1000)	(80) 100–600 (1200)	µg Zn/L
Copper	(2.0) 4.0–15 (20)	(2) 5–30 (60)	µg Cu/L
Iron	(< 10) 10–25 (50)	(50) 100–150 (300)	µg Fe/L
Selenium	(15) 20–45 (75)	(2) 5–20 (50)	µg Se/L
Chromium	< 1.0–1.5	(0.05) 0.1–1.5 (10)	µg Cr/L
Manganese	< 1.0	(0.08) 0.1–3	µg Mn/L
Aluminum	< 5–25	3–30	µg Al/L
Lead	(0.35) 1.0–10.0 (12)	(0.5) 4–20 (45)	µg Pb/L
Cadmium	< 0.5–1.0	(0.02) 0.2–1.0 (3)	µg Cd/L
Mercury	< 0.1–2.0	0.1–2 (20)	µg Hg/L
Thallium	< 0.1–2.0	(0.1) 1.0–1.5 (20)	µg Tl/L
Arsenic	< 1–30	3–10 (25)	µg As/L
Strontium	(20) 30–250 (350)	(< 10) 30–100	µg Sr/L
Barium	(0.5) 1.5–5.0 (10.0)	≈ 5	µg Ba/L

Table 6. Reference ranges for trace elements in scalp hair

Element	ESB results	Data from literature	Unit
Zinc	(30) 50–300 (500)	(40) 55–250 (450)	µg Zn/g
Copper	(15) 20–100 (200)	(2) 5–45 (150)	µg Cu/g
Iron	(5) 10–35 (50)	5–45 (150)	µg Fe/g
Manganese	(0.1) 0.3–1.0 (2.0)	(0.03) 0.1–1.5 (5.5)	µg Mn/g
Chromium	(0.01) 0.1–5.0 (20)	(0.02) 0.1–3 (15)	µg Cr/g
Nickel	(< 0.1) 0.1–3.0 (8.0)	(0.001) 0.04–1.25 (10)	µg Ni/g
Selenium	(0.50) 0.90–2.0 (2.5)	(0.02) 0.2–2.5	µg Se/g
Aluminum	1.0–20	1–20 (30)	µg Al/g
Strontium	(0.5) 1.0–5.0 (12)	(0.05) 0.1–1.0 (15)	µg Sr/g
Barium	0.20–1.0	0.1–1.0 (5.0)	µg Ba/g
Lead	(0.05) 0.1–2.0 (5.0)	(0.1) 0.2–15 (80)	µg Pb/g
Cadmium	(0.02) 0.05–0.25 (0.3)	(0.08) 0.25–2.0 (5.0)	µg Cd/g
Mercury	(< 0.01) 0.01–8.0 (15)	(0.05) 0.2–2.5 (10)	µg Hg/g
Arsenic	0.005–0.5	(0.01) 0.1–0.5 (1.5)	µg As/g

true for whole blood, human milk and adipose tissue. For inorganic trace compounds the correlations are not yet known enough to allow environmental exposure tests with only one specimen. Table 7 summarizes a selection of the known trace element toxicokinetic indicator matrices. 'Possible' indicator samples require further detailed research and comparison with data from autopsy material. Toxicodynamic indicators may be used to ensure exposure history.

Some Practical Results and Examples for Human Biomonitoring

Acute Exposure Monitoring

Due to the high sensitivity of the plant species *Vinis vitifera* to numerous pests, wine growers may be exposed to several insecticides and fungicides. Figure 6

Table 7. Elementspecific indicator specimens; toxicokinetic and toxicodynamic indicators

Element	Toxicokinetic indicators 'Available' samples Indicator function		Tissue	Toxicodynamic indicators
	ensured	possible		
Cadmium	urine/whole blood	hair	liver/kidney	β_2-microglobulin in urine/total protein in urine
Lead	whole blood/urine	hair	bone	ALA-D/FEP/δ-amino-levulinic acid in urine/nerve conduction
Arsenic	urine/hair/nails		skin/bone	
Mercury	urine	whole blood/hair	liver	lysosomal enzymes in serum/protein in urine/neuromuscular function
Thallium	urine/hair	feces/whole blood	liver/kidney	microscopic examination of hair root
Platinum	urine/blood plasma	hair		
Aluminum	blood plasma	urine/cerebro-spinal fluid	bone	P/Ca/alkaline phosphatase in serum
Zinc	urine/blood plasma/hair	saliva	bone	retinol binding protein/taste and smell test
Copper	blood plasma/hair	urine	liver/brain	ceruloplasmin in serum
Chromium	urine/hair/red blood cells			glucose tolerance/hyaluronidase in serum
Molybdenum	urine/blood plasma/hair		liver/kidney/spleen	ceruloplasmin in serum/Cu in urine/serum and urine uric acid/xanthinoxidase activity in erythrocytes
Selenium	urine/hair/blood plasma			glutathione peroxidase in red blood cells

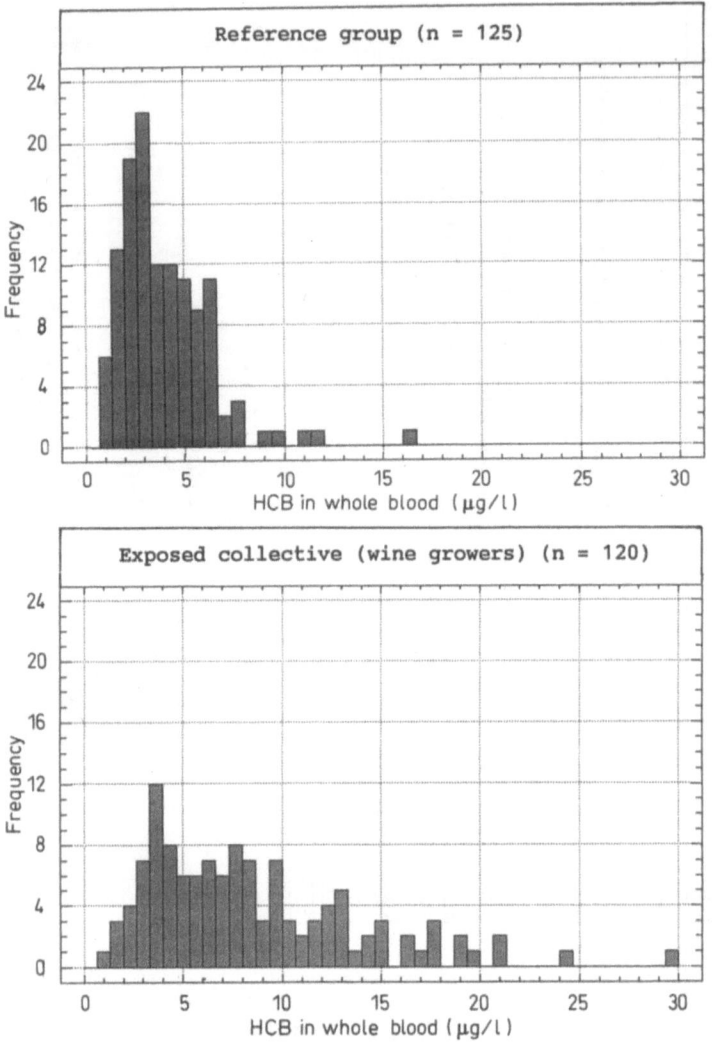

Fig. 6. Distribution of hexachlorobenzene in whole blood in a reference group (n = 125) (above) and a pesticide exposed group (wine growers) (n = 120) (below) (from [1])

demonstrates the increased body burden of hexachlorobenzene (HCB) in inhabitants of a small village in the Moselle valley, compared with the distribution of a control group. In the same wine growers the exposure to one of the most used inorganic fungicide (copper compounds) could be monitored in blood plasma (Table 8).

Information about the specific exposure hazard of people living in the surroundings of plants with probable toxic emissions may be revealed by monitoring an 'elemental scan'. Figure 7 demonstrates the individual 'metal

Table 8. Copper content in blood plasma in an exposed group (wine growers) compared with a reference group (mg Cu/l)

	Exposed group	Reference group
n	122	342
range	0.590–2.078	0.271–2.170
quantiles		
50%	1.134	0.830
75%	1.306	0.990
85%	1.369	1.117
95%	1.709	1.378
median	1.134	0.830
average	1.160	0.877
standard deviation	0.272	0.257
variance	0.074	0.066

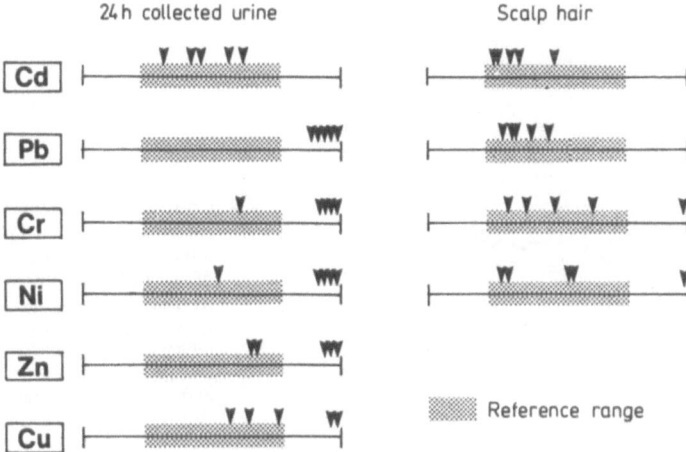

Fig. 7. Individual 'metal fingerprints' in 24-h-collected urine and scalp hair of 5 humans living in the immission area of an enameling factory

fingerprints' of 5 people living in the immission area of an enemaling factory. The elements analyzed were selected according to the heavy metal content in the household dust in this area (Table 9). The useful role of urine as a biomonitoring specimen in contrast to the scalp hair is evident. Obviously marginal environmental exposure to cadmium is indicated neither by urine nor by hair. As mentioned in Table 7 toxicodynamic indicators (e.g. specific proteins in urine) must be included in cadmium monitoring, especially in people with an elevated Cd content as a consequence of occupational exposure.

Risk assessment in the field of health protection has to be based on reliable epidemiological results. Biomonitoring may contribute sufficient data pools to

Table 9. Composition of household dust in the immission area of an emaille plant (mg/kg)

Element	Content
Chromium	85–210
Nickel	2250
Lead	300–800
Cadmium	5–8
Zinc	450–500
Copper	230

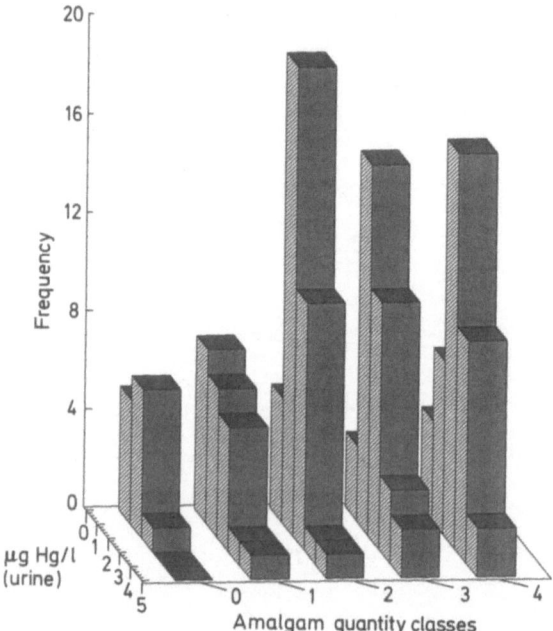

Fig. 8. Correlation of mercury contents in urine to the number of amalgam fillings (amalgam quantity classes: 0 = no amalgam/1 = 1–2 amalgam fillings/2 = 3–5 amalgam fillings/3 = 6–10 amalgam fillings/4 = more than 10 amalgam fillings)

the discussion of specific problems: e.g. silver amalgam is widely used as dental filling material. Mercury liberation and systemic Hg storage is discussed as a potential health risk. Biomonitoring results showed a correlation of mercury content in the indicator specimen urine to the quantity of amalgam fillings (Fig. 8). Calculation revealed in fact a minor risk of Hg exposure: Mercury resorption from this source was in the same order of Hg ingestion with food and distinctly lower than the 'provisional tolerable weekly intake', recommended by WHO/FAO (300 μg Hg). The other elements in this mercury alloy (silver, tin, copper) are of no toxicological importance.

Trend Analysis

Pentachlorophenol (PCP) was widely used as indoor wood protective until its production and use were restricted in 1978 and stopped completely in 1987

(FRG). During the years 1983–1989 biomonitoring of blood plasma and urine showed a continuous decrease of PCP levels. This drop corresponds to the decrease of hexachlorobenzene (HCB) plasma levels (Fig. 9). Since PCP is not a persistent xenobiotic in the human body, the basic level of this compound in urine and plasma without any further PCP exposure may be the result of HCB biotransformation as it was shown in our laboratory [7].

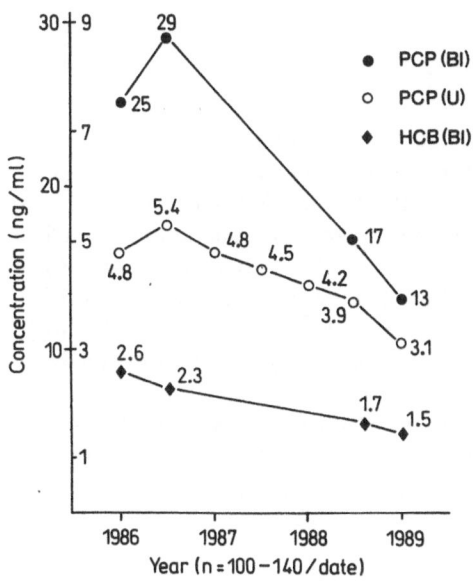

Fig. 9. Decrease in the hexachlorobenzene (HCB) and pentachlorophenol (PCP) levels in blood (Bl) and urine (U) during the period 1983–1989 (median)

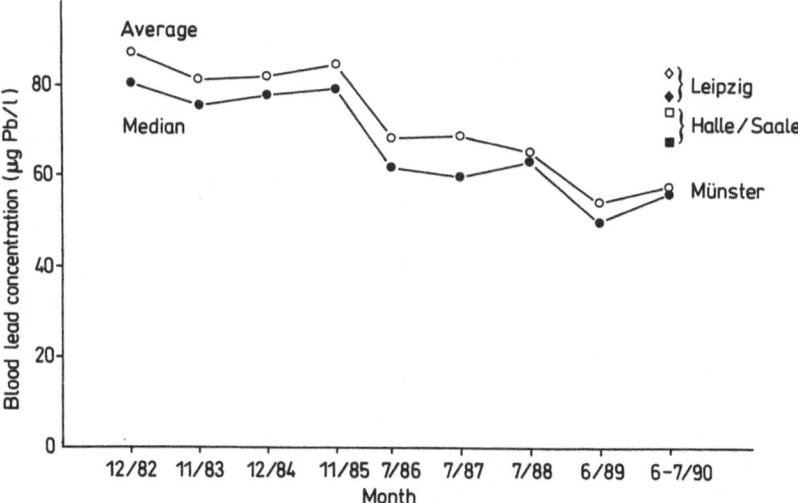

Fig. 10. Time course of lead in whole blood 1982–1990

Similar observations were found, when the content of lead in blood was biomonitored from 1982–1990. A distinct drop from 1984 on could be detected, which correlated to the legislative decrease of lead in gasoline. Thus biomonitoring may not only supervise the efficiacy of environmental protection steps, but also allows comparison of 'reference ranges' due to different environmental exposure: In Fig. 10 the reduction of the lead level in blood during 1982–1990 in FRG is given and compared with two other areas (1990) of obviously higher Pb exposure (former GDR).

Conclusion

It can be stated that carefully selected human organ tissue and body fluids are reliable sources for biomonitoring if standardized and well evaluated analytical methods are applied. Specimens of human origin can also be characterized by a thorough history of the donor and thus have a "mirror"-function as related to exposition and body burden, as well as to kinetic and dynamic properties of xenobiotic factors. Though the observation time period of human organ tissue sampling and banking is relatively short, it has already been proven for some important organic and inorganic pollutants that their changes in the environment can be detected in an unexpected short time.

Thus specimens of human origin play an important role in biomonitoring and in completing the knowledge about the actual (RTM) contents of xenobiotics as well as after long lasting exposure. They are tools in legal decision making processes and in trend analyses.

References

1. Bertram HP, Kemper FH, Zenzen C (1985) Man – a Target of ecotoxicological influences. In: Nürnberg HW (ed) Pollutants and their ecotoxicological significance. John Wiley, Chichester, p 415
2. Kemper FH, Bertram HP, Eckard R, Müller C (1988) Monitoring und Lagerung von Human-Organproben. In: Bundesministerium für Forschung und Technologie (ed) Umweltprobenbank. Springer, Berlin Heidelberg New York, p 113
3. Krause C, Chutsch M (eds) (1987) Haaranalyse in Medizin und Umwelt. Gustav Fischer, Stuttgart
4. Jockenhövel F, Bals-Pratsch M, Bertram HP, Nieschlag E (1990) Seminal lead and copper in fertile and infertile men. Andrologia 22: 503
5. Bertram HP, Kemper FH, Müller C (1986) Hexachlorobenzene content in human whole blood and adipose tissue: Experiences in environmental specimen banking. In: Morris CR, Cabral JRP (eds) Hexachlorobenzene: Proceedings of an international symposium. International Agency for Research on Cancer, Lyon, p 173
6. Bertram HP, Robbers J, Schmidt R (1984) Multielementanalyse mit ET-AAS im Rahmen der Human-Umweltprobenbank Münster. Fresenius Z Anal Chem 317: 462
7. Eckard R, Kemper FH (1990) Normal Ranges and Trends of Human Xenobiotic Burden: Pentachlorophenol (PCP). In: Bulletin de la Société des Sciences Médicales du Grand-Duché de Luxembourg: Proceedings of the International Congress on Clinical Toxicology, Poison Control and Analytical Toxicology, LUX TOX'90, p 320

3 Specimen Banking in Industrialized Countries

3.1 Progress of Six Years Experience with Environmental Specimen Banking in the Federal Republic of Germany

U.R. Boehringer

The German Environmental Specimen Bank is discussed in this chapter. The selection of specimens and sampling areas is described and a short description of the tasks of the cooperating institutions is given.

Introduction

The idea of Frederick Coulston and Friedhelm Korte, as mentioned in the introduction of this special edition, to set up an environmental tissue bank was put to practice in the Federal Republic of Germany in 1975. Soon it was evident that limiting the collection only to human tissues [1] would not give a comprehensive picture of the environment and the degree of pollution. Therefore, it was decided to include collection and analyses of species of fauna and flora. It was also realized that it would not be possible to store all species or tissues of species which were of interest [2].

The ecologists considered about 100 different species for the specimen bank. The chemists explained that they could not handle such a work load; neither manpower nor resources were available to the required extent. Therefore, there was a consensus to set up a medium size Environmental Specimen Bank [3].

After some tentative R + D projects a comprehensive pilot study was started in 1978, mainly supported by the Federal Minister for Research and Technology, under the supervision of the Umweltbundesamt (Federal Environmental Agency). It was a technical feasibility study. The sampling of different species, the handling and shipping of samples, deep freezing, homogenization, ultra trace analyses, amounts for repeated analyses, packing materials, logistics, storage temperature, and documentation were studied [4]. The results of the pilot study were so encouraging that the Government of the Federal Republic of Germany decided to set up a permanent Environmental Specimen Bank.

Selection of Specimens

The two most important factors were the selection of suitable species and the sampling areas. Based on the experience of the pilot phase additional selection of

Specimen Banking
Rossbach/Schladot/Ostapczuk (Eds.)
© Springer-Verlag Berlin Heidelberg 1992

Table 1. Preliminary specimen types   (as of Dec. 1990)

Specimen Type	Latin Name	Specification	Remarks
1.–3. Soils			up to three horizons each.
4. Freshwater Sediment			not included for the time being
5. Marine Sediment			
6. Air ingredients			
7. Spruce	*Picea abies*	one year old shoots	
8. Poplar/Beech	*Populus nigra italica*	leaves	
	Fagus silvaticus	leaves	
9. Earthworm	*Allolobophora longa*		
	and Lumbricus rubellus		
10. Roe dear	*Capreolus capreolus*	♀ liver	
11. Pigeon	*Columba livia*	egg	specimen type
12. Herring gull	*Larus argentatus*	egg	at the Canadian Wildlife Service
13. Zebra mussel	*Dreissena polymorpha*	soft parts	
14. Common mussel	*Mytilus edulis*	soft parts	
15. Brown algae	*Fucus vesiculosus*		
16. Cod	*Gadus morrhua*	liver	pelagial
17. Flounder	*Platichthys flesus*	muscles	benthic
18. Bream	*Abramis brama*	muscles and other tissues	
19. Sewage sludge			
20. Human blood		Whole blood and plasma	
21. Human liver/kidney			
22. Human adipose tissue			
23. Human urine			
24. Human hair (RTM)			
25. Mother's milk			
(26. Darnel	*Lolium multiflorum*	leaves	deferred)

specimens had to be made. It included air ingredients, soil and sediment, as well as the species or tissues of various animals and plants.

The list of possible species is endless, but in the beginning the number of species was limited to 25 because of limited resources and limited space for appropriate storage. The list is considered a minimum by ecologists. After the pilot phase the chemists stated that only 10 different types of specimen could be processed and analyzed with the given manpower and budget. A compromise of 25 different specimens, but collected only every other year, was reached. Furthermore, it was determined that samples of soil or sediment must be sampled at intervals of not less than five years.

Table 1 shows the specimens which were collected, analyzed and stored in the Environmental Specimen Bank. Each new type of specimen required additional investigation. Although one may use previously accepted methods of collection and analyses, one must show that these methods are appropriate for the new specimen type. Unexpected difficulties appeared with the new specimens which took considerable time and effort to overcome.

In order to standardize the various procedures for collecting, storing, analyzing, etc., the various specimens, Standard Operating Procedures (SOPs) had to be developed. To date SOPs have been issued for 14 specimens.

Selection of Areas

Resources were not allocated for the selection of sampling areas until the feasibility study showed that a specimen bank was practical and useful. Again R + D projects had to be started.

Two different academic schools of thought worked on the selection process. One school suggested a list of regionally representative areas. It was based on the multivariate analysis of comprehensive areal data derived from the digitaliz- ation of a series of ecology-related maps. The selection followed the principles of representativeness and involved a broad set of multivariate procedures, which permits a maximum extrapolation of results obtained in individual study areas [5].

The other school of thought suggested a network of ecological assessment parks. This network included the principal ecosystems. It had a minimum number of sites that collectively provided a representative cross section of the major types of ecosystems that occur in the Federal Republic of Germany. The criteria of selection were of a more practical nature, e.g. representiveness of major ecosystems, productive and useful short-term research, protection and accessibility for research purposes.

Within selected ecosystems, the location of suitable areas for banking had to be found. One main criteria was the watershed as the natural unit of study in the terrestrial sphere.

The selection was the compilation of a rather large list of candidate areas. All areas should fit the minimum requirements for long-term ecosystem research,

Federal Minister for the
Environment, Nature
Conservation and Nuclear
Safety

Scientific
supervision

Federal Environmental Agency

– scientific and admin.   – supervision of R + D projects
  management and control   – secretariat of the Scientific
– further development   Advisory Council
– harmonization with national   – environmental information and
  and international activities   documentation system UMPLIS

Research Centre Jülich

Bank for biological samples

– collection of marine
  samples
– inorganic analysis
– logistics and transport

Münster University

Bank for human-organ samples

– human-organ samples
– analysis
– data processing for
  the specimen bank

**University of the
Saarland**

sampling

GBF Neuherberg

organic analysis

**Biochem. Institute,
Großhansdorf**

analysis for PAHs

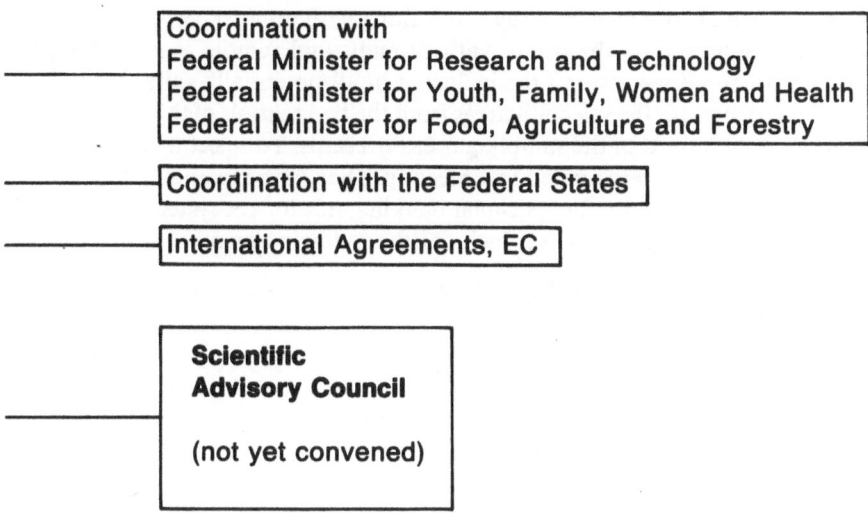

Fig. 1. Organizational scheme of the German Environmental Specimen Bank

monitoring and specimen banking. Consulting with other experts, reviewing the appropriate literature, the use of various maps and remote sensing data, and ground observations assured the inclusion of all potentially suitable sites [6].

If one compares the selection areas from the two schools of thought one finds that about two third of the sampling areas overlaps. To select a comprehensive network, scientists were invited to a meeting. They gave recommendations for a set of 14 areas, but only for specimen banking, not for ecosystem research and not for monitoring [7]. The set includes the following areas:

- North- and East-Frisian Wadden-Sea;
- Hilly country of Eastern Holstein;
- Solling (Lower Saxony);
- Bavarian Forest National Park;
- Hilly country between the Isar and Inn rivers;
- Berchtesgaden National Park;
- Lake Constance;
- Sections of the Rhine including Kühlkopf-Knoblochsaue (flood plain);
- Swabian Jura;
- Palatinate Forest;
- Saarbrücken;
- Frankfurt-on-Main;
- Berlin (West);
- Transect Aachen – Teutoburg Forest.

After selecting the areas, suitable sampling sites within these areas had to be chosen. This work is presently done in five areas. The selection of sampling sites in the other areas will begin at a later date.

Work Distribution and Organization

The Umweltbundesamt (Federal Environmental Agency) controls the Environmental Specimen Bank under the supervision of the Federal Minister for the Environment, Nature Conservation and Nuclear Safety. It is the responsibility of the Umweltbundesamt to manage the program, both scientifically and administratively, and also to coordinate it with the activities of other national and international institutions. The initiation of supplementary research is also a task of the Umweltbundesamt.

The work is distributed among five scientific institutions. Each of them has agreed to share the work depending on their special scientific capabilities. These five institutions are given in Fig. 1 [8]:

- The Research Center Jülich stores the specimens – except for human specimens – over liquid nitrogen in large vessels at minus 150 °C or cooler. It homogenizes the samples, analyzes them for inorganic ingredients, organizes the transport of all samples from the sampling sites to the bank or the laboratories and collects the marine samples.

- The University of the Saarland, Saarbrücken, collects the terrestrial and limnic samples, characterizes them and prepares them for transportation to the storage place.
- The Forschungszentrum für Umwelt und Gesundheit (GSF), Munich, analyzes the samples for halogenated compounds.
- The Biochemische Institut für Umweltcarcinogene, Großhansdorf (near Hamburg), analyzes the samples for polycyclic aromatic hydrocarbons (PAH). Not only the pollution level caused by PAH is determined by the analysis, but also the quality of sampling and handling of the specimens, because the occurrence of PAH in certain samples results from contamination due to handling.
- In Münster, the Westfälische Wilhelms-Universität performs all the activities with human samples, i.e. from collection to characterization, analysis and storage. The storage is carried out in a walk-in deep freezer at minus 80 °C. The database for all specimens is also in Münster, a geographical information systems will be included.

Outlook

Statements on pollution trends cannot be made at the present time. It will take at least ten years – that means five sampling periods – to obtain meaningful results for trend analyses. The natural variations are too great for an accurate assessment in a shorter time. Earlier statements about trends are considered to be educated guesses and not sufficiently accurate for making decisions.

The recent developments in Germany requires expansion of this program. The scientists for the eastern part of Germany have put forward ideas for the sampling sites. The sampling has to start as soon as possible to document the situation in the heavily polluted areas and to initiate remedial action and assess the progress of the relief projects.

This archiving and other work associated with the Environmental Specimen Bank will benefit future generations in Germany and in the rest of the world; we cooperate and coordinate our results with the other international Environmental Specimen Banks.

Acknowledgement: I wish to thank Dr. Frode Ulvedal, my former colleague at the U.S. Environmental Protection Agency, for reading this manuscript.

References

1. Berlin A, Wolff AH, Hasegawa Y (eds) (1979) The use of biological specimens for the assessment of human exposure to environmental pollutants. Martinus Nijhoff, The Hague
2. Luepke NP (ed) (1979) Monitoring environmental materials and specimen banking. Martinus Nijhoff, The Hague
3. Lewis AL, Stein N, Lewis CW (eds) (1984) Environmental specimen banking and monitoring as related to banking. Martinus Nijhoff, Boston

4. Bundesministerium für Forschung und Technologie (ed) (1988) Umweltprobenbank – Bericht und Bewertung der Pilotphase. Springer, Berlin Heidelberg New York
5. Fränzle O, Kuhnt D, Kuhnt G, Zölitz R (1987) Auswahl der Hauptforschungsräume für das Ökosystemforschungs programm der Bundesrepublik Deutschland. Umweltforschungsplan des Bundesministers für Umwelt, Naturschutz und Reaktorsicherheit, UFOPLAN-Nr. 101 04 043/02. Geographisches Institut der Universität Kiel
6. Lewis RA, Paulus M, Horras C, Klein B (1989) Auswahl und Empfehlung von ökologischen Umweltbeobachtungsgebieten in der Bundesrepublik Deutschland. In: Deutsches Nationalkomitee für das UNESCO-Programm "Der Mensch und die Biosphäre" MAB-Mitteilungen 29, Bonn
7. Der Rat von Sachverständigen für Umweltfragen (1990) Sondergutachten: Allgemeine ökologische Umweltbeobachtung. Wiesbaden
8. Boehringer UR, Hertel W (1987) Umweltproben unzersetzt langfristig lagern. Umwelt VDI-Verlag 1–2: 21

3.2 Specimen Banking at the National Institute of Standards and Technology

R. Zeisler[1], B.J. Koster, and S.A. Wise

More than twelve years of practical experience in specimen banking at the National Institute of Standards and Technology (NIST), within the National Biomonitoring Specimen Bank (NBSB), have demonstrated that the concept of long-term storage of environmental specimens is feasible. The activities at NIST include specimen banking of: human liver samples, samples from the marine environment (sediments, oysters mussels, fish tissue and marine mammal tissues), human serum, and total human diet samples. These research projects are associated with several different U.S. government environmental programs. The NBSB is providing a wide range of know-how in the collection, processing, long-term storage, and analysis of the different samples types. Even though the types of specimens and the number of samples collected are limited, the NBSB can serve as a valuable resource for the assessment of long-term trends of pollutants and for future retrospective studies. Specimens can be made available to the scientific community and national or international organizations.

Program Overview

For the past 12 years, the National Institute of Standards and Technology (NIST) has been involved in biological environmental specimen banking. In recent years, these activities at NIST have expanded beyond the pilot program with human livers to include samples from the marine environment (sediments, oysters, mussels, and fish tissue), marine mammal tissues, human serum, and total human diet samples. These research projects are associated with several different U.S. government programs, which are listed in Table 1. These projects are known collectively as the National Biomonitoring Specimen Bank (NBSB). The title reflects the inclusion of specimens of nutritional and medical importance in addition to those specimens archived for environmental purposes. The NBSB has provided a wide range of experience in the collection, processing, long-term storage, and analysis of different samples types. The projects are described briefly below, and a summary of the current inventory of the specimens included in the NBSB is provided in Table 2.

Pilot Environmental Specimen Bank

The goals and the development of the initial EPA/NIST pilot Environmental Specimen Bank Program have been described in detail in a previous publication [1], and the results and experiences of the early years of this project were

Specimen Banking
Rossbach/Schladot/Ostapczuk (Eds.)
© Springer-Verlag Berlin Heidelberg 1992

Table 1. Projects within the National Biomonitoring Specimen Bank at the National Institute of Standards and Technology

Project	Sponsoring organization/agency	Specimen type	Analytes of interest
Environmental Specimen Bank	Environmental Protection Agency (EPA)	Human liver Mussels	Trace elements, PCBs, PAHs, and pesticides
National Status and Trends Specimen Bank Project	National Oceanic and Atmospheric Administration (NOAA)	Mussels and oysters Sediment Fish muscle and liver	Trace elements, PCBs, PAHs, and pesticides
Alaskan Marine Mammal Tissue Archival Project	National Oceanic and Atmospheric Administration (NOAA) Mineral Management Services (MMS)	Marine mammal tissues (blubber, kidney, and liver)	Trace elements, PCBs, PAHs, and pesticides
National Marine Mammal Tissue Bank	National Oceanic and Atmospheric Administration (NOAA)	Marine mammal tissues (blubber and liver)	
Cancer Chemoprevention Program	National Cancer Institute (NCI)	Human serum	Organic nutrients (e.g., vitamins)
Trace Nutrients in Human Diet Project	International Atomic Energy Agency (IAEA) U.S. Department of Agriculture (USDA) Food and Drug Administration (FDA)	Total human diet	Inorganic nutrients

Table 2. Inventory of Specimens in the National Biomonitoring Specimen Bank

Human Specimens	
Liver (EPA)	559 specimens
Serum (NCI)	452 samples
Marine Specimens	
Mussels (EPA)	93 batches (70/batch) from one site
Benthic Surveillance Project (NOAA)	
Fish liver/muscle	26 sites
Sediments	29 sites
Mussel Watch Project (NOAA)	
Mussels/oysters	160 sites
Sediments	138 sites
Alaska Marine Mammal Tissues Archive Project (NOAA/MMS)	
Northern Fur Seal	15 animals
Ringed Seal	15 animals
Bearded Seal	3 animals
Harbor Seal	3 animals
Stellar Sea Lion	1 animal
Belukha Whale	14 animals
National Marine Mammal Tissue Bank (NOAA)	
Harbor Porpoise	1 animal
Pilot Whale	5 animals
Food Specimens	
Total Human Diet	8 collections
(IAEA/FDA/USDA)	

reviewed in 1984 [2]. This pilot effort was originally intended to focus on four types of environmental specimens: human liver, marine specimens, food specimens, and an atmospheric accumulator. However, since other government agencies involved in projects related to the marine environment and food specimens have joined the NBSB program, the EPA/NIST pilot specimen bank project has focused primarily on the establishment of a human liver bank and research related to specimen banking. The pilot phase ended with the establishment of an initial database on 96 stored human liver specimens (see discussion below). An additional 450 not yet analyzed liver specimens in the bank represent a substantial resource for the sponsoring agency, EPA, and other research projects for the investigation of environmental pollution trends. The human liver project continues with current collections and analyses.

National Status and Trends Specimen Bank

In 1985 the National Oceanic and Atmospheric Administration (NOAA) incorporated specimen banking into their National Status and Trends (NS&T) Program. The NS&T program is a national monitoring program designed to quantify the current status and long-term trends in the concentration of selected contaminants and biological indicators of contaminant effects in U.S. coastal and estuarine environments [3]. The NS&T program consists of two separate monitoring activities: (1) the Benthic Surveillance Project, in which sediment and fish tissues (muscle and liver) are collected annually from approximately 50 coastal sites, and (2) the Mussel Watch Project, in which sediment and bivalve molluscs (mussels and oysters) are collected from nearly 200 coastal sites. Each year samples from approximately 15% of these sites are collected specifically for archiving in the specimen bank; thus, samples from almost all Benthic Surveillance and Mussel Watch sites have been archived since the start of this project. Approximately 10% of all sites have been extensively characterized with the advanced analytical approaches of the NBSB [4].

Alaskan Marine Mammal Tissue Archival Project

The Alaskan Marine Mammal Tissue Archival Project (AMMTAP) was initiated in 1987 with the goal of establishing a representative collection of tissues from Alaskan marine mammals (e.g. seals, walruses, and whales) for future contaminant analyses and documentation of long-term trends in environmental quality. Since most marine mammals are at or near the top of the food chain, chemical analysis of their tissues may be useful in determining whether bioaccumulation of contaminants associated with human industrial activities is occurring in the marine food chains in the Arctic. In addition, some of the native population of Alaska depend upon such animals for a substantial portion of their diet. Therefore, the contaminant levels found in marine mammals may have health implications for the human population occupying these regions. A detailed discussion of the project including the rationale for the selection of specimens and the sample collection protocols has been published [5]. A small number of tissues from various species have been characterized extensively.

National Marine Mammal Tissue Bank

Increasing public and scientific interest in the health of marine mammals and the successful implementation of the AMMTAP has made it desirable to expand the latter program to all U.S. coastal waters (and beyond) in a National Marine Mammal Tissue Bank (NMMTB). In 1989, a project was initiated with NOAA to develop the strategy for the establishment of the NMMTB. The feasibility of specimen acquisition for the NMMTB from strandings and from incidental catches of marine mammals during commercial fishing operations is being explored. Each year hundreds of marine mammals are found stranded on our coasts, and other marine mammals are taken incidentally during commerical fishing operations. In the case of strandings, regional marine mammal stranding networks currently exist to collect information on stranded animals. Under appropriate conditions, such stranding networks will be used for collecting tissue samples for the NMMTB. Tissue samples will also be collected from marine mammals taken incidentally during commercial fishing operations. The Fisheries Conservation and Management Act (FCMA) requires observers to be on board all foreign fishing vessels operating in U.S. waters. Under the 1988 Amendments to the Marine Mammal Protection Act (MMPA), observers will be required to be on various U.S. fishing vessels. These observers, under the authority of NOAA National Marine Fisheries Service, will be instructed to identify suitable animals for the NMMTB from marine mammals taken incidentally.

In 1990 a demonstration phase was established to evaluate the practical aspects of obtaining suitable specimens for the NMMTB from both incidental catches and strandings. Specimens were obtained from harbor porpoises (*Phocoena phocoena*) taken incidentally and pilot whales (*Globicephala macrorhyncha*) from a mass stranding. The objectives for collection, the collection protocols, and the banking procedures are similar to the AMMTAP. A significant outcome of the initial phase of this project will be a "stranding handbook" that will detail all procedures necessary for obtaining valid samples for chemical and other investigations and for banking.

NCI Cancer Chemoprevention Program and IAEA/USDA/FDA Nutrients in Human Diet

Two additional projects, the Cancer Chemoprevention Program and the Nutrients in Human Diet Project, are not specifically directed toward specimen banking as are the EPA, MMS, and NOAA projects; however, they do have minor specimen banking components associated with them. Both of these projects focus on nutrients rather than on environmental contaminants as in the four major projects described above. As part of a quality assurance program for the NCI's Cancer Chemoprevention Program, NIST serves as a reference laboratory for NCI-supported laboratories involved in the determination of micronutrients in human serum (e.g. vitamins A, C, and E and β-carotene). Large batches of human serum samples containing measured amounts of these analytes are prepared and distributed as proficiency testing samples for the various laboratories. Since the long-term stability of these nutrients in serum is unknown, selected well-characterized specimens are stored under various conditions to determine storage stability. The Nutrients in Human Diet Project is a joint program among several different agencies to obtain comparative data on dietary intakes of nutritionally important minor and trace elements in a number of countries [6]. As part of this effort, samples of the total human diet composites, collected as part of the FDA "market basket" survey [7], are banked for long-term storage.

Program Accomplishments

Development of Procedures/Techniques for Noncontaminating Collection and Processing of Biological Samples

Sample Collection Procedures

The purpose of a specimen bank is the long-term preservation of specimens that are representative of the state of a site or organism immediately prior to collection. Therefore, a major concern of specimen banking efforts is that the samples be collected, processed, and stored under conditions that avoid or minimize contamination or any other changes in the chemical composition of the specimen. In most instances, environmental samples contain extremely low levels of inorganic and organic contaminants; therefore, extreme caution must be exercised to ensure that the samples are not contaminated. Detailed sample collection protocols have been developed and implemented for each of the specimen types in the various projects, i.e. human liver [1]; sediments, fish tissue, and mussels/oysters [8]; and marine mammal tissues [5].

The basic philosophy for the development of these protocols focuses on the use of noncontaminating materials whenever there is any contact with the sample. For example, when samples require dissection (e.g. in the case of human liver, marine mammal, or fish tissues), a titanium-bladed knife is used to avoid contamination from environmentally important trace elements (e.g. Ni, Cr, and Fe) found in conventional cutting instruments. During sample preparation, contact with the specimen is generally limited to clean, dust-free Teflon surfaces and the specimens are stored in Teflon bags or jars. After the specimens are placed in the storage containers, they are frozen in liquid nitrogen as soon as feasible and transported to the specimen bank facility at NIST, where the samples are stored in liquid nitrogen vapor freezers at $-150\,°C$. The specimens are preserved at this temperature until the time of analysis. Each of the collection protocols for the different specimen types was developed, in conjunction with individuals involved in the sampling, to achieve a suitable non-contaminating procedure within the bounds of practicality. As part of the sampling protocol, information describing the sample and the sampling site are recorded; this information is maintained, both in hard copy and in a computer database, as part of the documentation for each sample in the specimen bank.

Cryogenic Homogenization

The preparation of homogeneous sample aliquots from the bulk sample is a major requirement for specimen banking. Identical (i.e. homogeneous) sample aliquots are necessary to allow for valid comparison of data obtained by various researchers and analytical techniques and for evaluation of the stability of specimens during storage. To address this requirement, a cryogenic homogenization procedure has been developed using Teflon disk mills [9]. These mills are capable of homogenizing 150-g sample aliquots to provide homogeneous frozen samples with greater than 90% of the particles less than 0.46 mm in diameter and with subsampling errors due to inhomogeneity estimated at less than 2%. Since the initial evaluation and report describing this procedure, it has been successfully used for the homogenization of a variety of specimen types including: human liver and adipose tissues, mussel and oyster tissues, fish tissues (liver and muscle), honey bees, marine mammal tissues (liver, kidney, muscle, and blubber), chicken tissue, and total human diet composites. The cryogenic procedure uses Teflon mills to minimize contamination and eliminates the risk of potential changes in the sample associated with thawing and refreezing. A similar approach has been used in the Environmental Specimen Bank program in Germany where a "continuous flow" apparatus has been developed for the preparation of large quantities of homogenous frozen material [10].

Chemical Characterization of Specimens

Baseline Environmental Analytical Data

As part of the specimen banking activities, approximately 10% of the archived specimens are analyzed to determine selected organic and inorganic constituents. These analyses provide accurate baseline data for the following purposes: (1) for use in evaluating the stability of the specimens during long-term storage, (2) for comparison with data obtained from other laboratories analyzing similar samples, which were collected at the same time from the same site as part of a monitoring program, (3) for comparison with data from samples to be collected in the future for monitoring long-term trends in pollution at a particular site, and (4) for comparison with data from different sites for real-time monitoring. An important aspect of our specimen banking activities is that they provide both inorganic and organic baseline data on the same specimens.

Human Livers: Of the 550 human liver specimens collected since 1980, 96 specimens have been analyzed to provide data on about 30 trace elements per specimen. These 96 specimens were analyzed in three groups of samples collected in 1980, 1982, and 1984. The 1980 samples were from three locations, i.e. Baltimore, MD; Minneapolis, MN; and Seattle, WA; whereas the 1982 and 1984 were primarily from Seattle. The inorganic data sets for the 96 samples from 1980, 1982 and 1984 have been published [1, 11, 12], and the complete results obtained by nuclear techniques are summarized in this volume [4]. In addition, organic analyses of aliquots from 50 of the same samples have been performed for the determination of chlorinated pesticides and polychlorinated biphenyls (PCBs) [13].

One observation from the inorganic baseline data for the human livers [1, 2] is that many of the pollutant trace element concentrations are on the low end of, or below, previously reported ranges [14]. Specifically, the levels of Al, As, Tl, Sn, and Pb are lower than previously reported concentrations of these elements in human liver from 1940–1972. More recent compilations of laboratory results [15] have included the data from this work and are in general agreement with our findings.

Marine Specimens: During the three years of collection and analysis of marine specimens as part of the NOAA NS&T program, sediment, fish muscle, and fish liver samples from 12 sites in the Benthic Surveillance Project and sediment and mussels/oysters from 18 sites in the Mussel Watch Project have been analyzed. Baseline data for approximately 30 trace elements in the fish muscle and liver samples and 45 trace elements in the mussel, oyster, and sediment samples have been determined in these 30 U.S. coastal sites [4]. Some pollutant elements that are not commonly determined, such as silver and vanadium, can give the opportunity to assess different pathways of environmental input, i.e. urban and petrogenic pollution. Organic analyses have provided data for selected polycyclic aromatic hydrocarbons (PAHs), poly-chlorinated biphenyls (PCBs, individual congeners), and chlorinated pesticides for samples from these same 30 sites [16]. Differences in 4,4'-DDE and 4,4'-DDT ratios give estimates on recent uses of this pesticide. From the Alaskan Marine Mammal Tissue Archival Project, muscle, blubber, kidney, and liver tissues from the 1987 collection of northern fur seal (*Callorhinus ursinus*) samples have been analyzed [17]. Similar measurements have been completed on tissue specimens from the 1988 collection of ringed seals (*Phoca hispida*) and the 1989 collection of belukha whales (*Delphinapterus leucas*) [18]. The determination of a large number of PCB congeners in different specimen types may lead to new information on accumulation and detoxification in various organs of the marine mammals.

Retrospective Determinations

An important justification for specimen banking is that analytical methods are continually improving, and thus the availability of banked samples allows researchers to apply these improved procedures to specimens from the past. During our ten years of banking human liver specimens, we have witnessed a number of such advances and/or improvements in analytical procedures. The determination of As in the human liver specimens is an example where the analytical procedures lacked sufficient sensitivity at the time the samples were initially collected. Based on the literature data available in 1980 for the concentration range of As in liver, i.e. from 11 to 460 μg/kg, atomic absorption spectroscopy (AAS) was used to analyze the samples. However, the concentrations of As in the samples were below the expected levels and very near the detection limit of the AAS technique. Since 1980 a more sensitive radiochemical neutron activation analysis (RNAA) procedure has been developed for the determination of As [12]. In our efforts to determine the stability of banked specimens (see discussion below), we reanalyzed the human liver specimens from the 1980 collection during 1988 using the RNAA procedure that was not available when the samples were originally analyzed in 1980. The results range from 1.5 to 21 μg/kg with a median of 6.1 μg/kg. No trend was observed among the three sets of human livers from 1980, 1982, and 1984.

An excellent opportunity for retrospective analyses exists for the determination of ultra-trace levels of Pt in banked human liver specimens. Recently, interest in the level of Pt in the environment has increased due to the use of Pt in the catalytic converters on many automobiles. In 1980 when the human liver samples were collected, sensitive methods capable of determining the low levels of Pt in human livers (10–70 pg/g) were not available. Since then methods for the accurate determination of these levels in liver have been developed and baseline data on 12 samples from the 1980 collection have been reported [19]. The opportunity now exists to apply these sensitive techniques to a significant number of samples that were collected since 1980 to verify whether Pt concentrations are increasing or decreasing in the human population.

Feasibility of Specimen Banking

Long Term Stability of Banked Samples

A critical requirement for a specimen bank program is the assurance that samples do not undergo changes during storage. Hence, one goal of the EPA pilot Environmental Specimen Bank project was to determine the stability of specimens under various conditions such as freeze-dried and stored at room temperature, and fresh frozen and stored at $-25\,°C$, $-80\,°C$, and $-150\,°C$. To address the question of storage stability, aliquots (ca. 6–8 g) of the homogenized human liver samples were stored under the above conditions. After seven years of storage, aliquots stored under two of the conditions ($-25\,°C$ and $-150\,°C$) were reanalyzed for comparison with baseline data obtained immediately after the homogenization in 1980 (for inorganic constituents only) and for direct comparison of the two storage conditions. The correlations between baseline data and data obtained after seven years of storage, for 20 elements in 24 samples, have been investigated. The inorganic results did not show any deviations. Correlation coefficients of r 0.97 have been observed for two storage conditions, at $-25\,°C$ and at $-150\,°C$ [20].

Aliquots of the same samples were also analyzed for pesticides and selected PCB congeners. Results for 4,4′-DDE and PCB 180 in 24 subsamples stored at $-25\,°C$ and $-150\,°C$ are shown in Fig. 1. As with the trace element results, no significant changes were observed between the samples stored at $-25\,°C$ and $-150\,°C$ as indicated by the slope of the regression lines. Unfortunately, no organic baseline data for the 1980 liver samples were available to determine whether analyte concentrations had changed at both of the storage conditions since the initial storage.

Even though the chemical analyses of the samples stored at $-25\,°C$ and $-150\,°C$ indicated no significant changes in composition, there was physical evidence of changes in the sample aliquots. At $-25\,°C$ the aliquots of frozen liver homogenate had formed ice crystals under the container lids and on the sample surface (i.e. the moisture in the samples had separated), and the homogenates were no longer powdery but were clumped. The samples stored at $-150\,°C$ were still powdery, as they had been at the time of homogenization. The subsample weights were stable over the storage interval for both storage conditions. The separation of moisture from the samples stored at $-25\,°C$ would necessitate the use of the total subsample for any analytical determinations. In addition, the color of the sample aliquots stored at $-25\,°C$ was different from those stored at $-150\,°C$. Samples stored at $-80\,°C$ did not show any formation of ice crystals or any noticeable color changes. Reanalyses

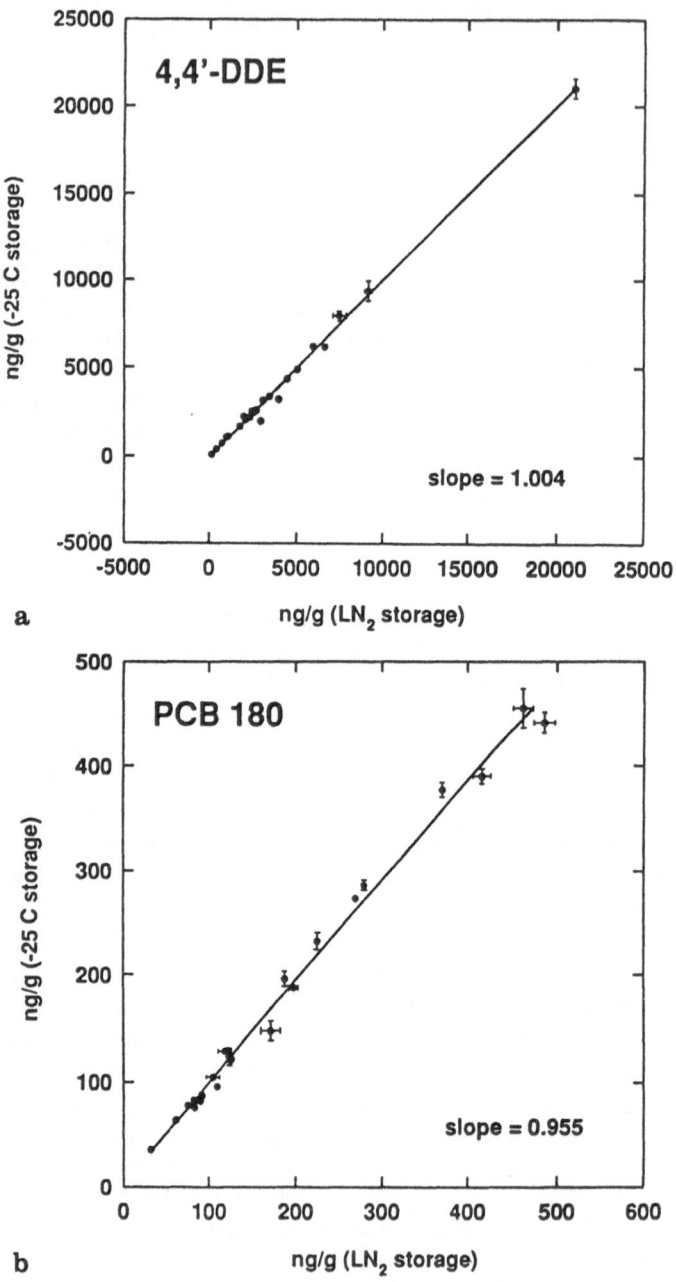

Fig. 1a, b. Correlation plot of results in human liver subsamples stored at − 25 °C and − 150 °C for seven years for the determination of a) 4,4′-DDE and b) PCB 180. Concentrations in μg/kg extractable fat of banked samples will continue in the NBSB projects to further assess the effects of long-term storage at different conditions. Currently all samples are banked at − 150 °C to avoid the physical changes noted above and because of the relative maintenance-free, low-cost operation of the liquid nitrogen vapor freezers.

Logistical and Organizational Aspects

Besides maintaining samples under the controlled conditions as discussed above, the NBSB is concerned with all aspects of collecting specimens, updating information, and of disemminating samples for testing purposes. The NBSB is maintaining close contacts with the scientists in the collaborating research and monitoring projects. Sampling protocols are developed and updated in coopera-tion with the scientists in the field, since the quality of logistic and support facilities available for the collections vary widely throughout the NBSB projects. Modifications to the collection procedures are made and documented as required to assure maximum protection of the samples even in remote locations. Completely packaged sampling kits contained in sturdy boxes, such as ice chests, designed for the collection of one or more specimens, have become very useful on board small boats or research vessels, and in improvised laboratories on shore. Each kit usually has all the protocol materials, protective clothing for the sampling personnel, and material to serve as clean working surfaces for sample preparation. The kits are supplied by the NBSB to avoid substitutions in the field. The additional cost in sampling, revision of protocols, and visits and discussions with field personnel is outweighed by the reliability of the specimens stored in the bank.

Since the NBSB is a repository of systematically collected, characterized and stored samples for deferred examinations, mechanisms are in place for outside investigators to obtain access to portions of the specimens. Typically, a proposal for use of a portion of banked samples will be reviewed by the individual project manager of the collaborative project in conjunction with the NBSB program management. At most, 50% of a specimen will be available for general investiga-tions (all specimens have been stored in duplicate portions). In practice, only small portions of the previously homogenized and generally characterized specimens are currently investigated by researchers outside the NBSB projects. The selection of samples has been limited to these specimens because it is necessary to minimize the expenses for homogenization and additional storage. Of course, these expenses can be covered by the prospective investigator. The NBSB would like to support as many requests as possible to more completely characterize the specimens in the bank. Through the additional investigations, the specimens will become even more valuable for retrospective investigations.

Conclusions

More than twelve years of practical experience in specimen banking within the National Biomonitoring Specimen Bank have demonstrated that the concept of long-term storage of environmental specimens is feasible. Although the total scientific value of the banked samples is not fully known at this time, the current uses of the banked samples and the implementation of the concept has already contributed to major monitoring programs in the U.S. and abroad. Even though the types of specimens and the number of samples collected are limited, the NBSB can serve as a valuable resource for the assessment of long-term trends of pollutants affecting human and environmental health, in particular for those pollutants that have been unnoticed up to now or that could not be measured in the past.

Acknowledgements: This work was supported in part by the Office of Health Research, Office of Research and Development of the Environmental Protection Agency; the Office of Oceanography and Marine Assessment, Ocean Assessments Division, National Ocean Service, and the Office of Protected Resources, National Marine Fisheries Service of the National Oceanic and Atmospheric Administration; and the Minerals Management Service, Department of Interior.

References

1. Zeisler R, Harrison SH, Wise SA (eds) (1983) The Pilot National Environmental Specimen Bank – Analysis of human liver specimens. Natl Bur Stds Spec Publ 656, US Government Printing Office, Washington DC 135 p
2. Wise SA, Zeisler R (1984) Environ Sci Technol 18: 302A
3. Lauenstein GG, Calder JA (1988) In: Wise SA, Zeisler R and Goldstein GM (eds) Progress in environmental specimen banking. Natl Bur Stds Spec Publ 740, US Government Printing Office, Washington DC 19 p
4. Stone SF, Zeisler R (1991) In this book
5. Becker PR, Wise SA, Koster BJ, Zeisler R (1988) Alaskan Marine Mammal Tissue Archival Project: A project description including protocols, NBSIR 88-3750, Natl Inst Stds Techn, Gaithersburg MD 53 pp
6. Iyengar GV, Tanner JT, Wolf WR, Zeisler R (1987) Sci Total Environ 61: 235
7. Pennington JAT, Young BE, Wilson DB, Johnson RD, Vanderveen JE (1986) J Am Diet Assoc 86: 876
8. Lauenstein GG, Wise SA, Zeisler R, Koster BJ, Schantz MM, Golembiewska SL (1987) National Status and Trends Program for Marine Environmental Quality Specimen Bank Project: Field Manual. NOAA Technical Memorandum NOS OMA 37, National Oceanic and Atmospheric Administration, Rockville MD 38 pp
9. Zeisler R, Langland JK, Harrison SH (1983) Anal Chem 55: 2431
10. Schladot JD, Backhaus FW (1988) In: Wise SA, Zeisler R, Goldstein GM (eds) Progress in environmental specimen banking. Natl Bur Stds Spec Publ 740, US Government Printing Office, Washington DC 184 p
11. Zeisler R, Harrison SH, Wise SA (1984) Biol Trace Element Res 6: 31
12. Zeisler R, Greenberg RR, Stone SF (1988) J Radioanal Nucl Chem 124: 47
13. Parris RM, Chesler SN, Wise SA, In: Wise SA, Zeisler R, Goldstein GM (eds) Progress in environmental specimen banking. Natl Bur Stds Spec Publ 740, US Government Printing Office, Washington DC 74 p

14. Iyengar GV, Kollmer WE, Bowen HJM (1978) The elemental composition of human tissues and body fluids. Verlag Chemie, Weinheim, Germany
15. Iyengar GV, Woittiez JRW (1988) Clin Chem 34: 474
16. Schantz MM, Chesler SN, Koster BJ, Wise SA (1988) In: Wise SA, Zeisler R, Goldstein GM (eds) Progress in environmental specimen banking. Natl Bur Stds Spec Publ 740, US Government Printing Office, Washington DC 40 p
17. Becker PR, Wise SA, Zeisler R (1989) Alaskan Marine Mammal Tissue Archival Project: Acquisition and curation of Alaskan marine mammal tissues for determining levels of contaminants associated with offshore oil and gas development. National Atmospheric and Oceanic Administration, Oceans Assessment Division, Anchorage AK 98 pp
18. Becker PR, Wise SA, Zeisler R (1991) Alaskan Marine Mammal Tissue Archival Project. National Atmospheric and Oceanic Administration, Oceans Assessment Division, Anchorage AK (in press)
19. Zeisler R, Greenberg RR (1988) In: Brätter P, Schramel P (eds) Trace element analytical chemistry in medicine and biology. Walter de Gruyter, Berlin, 297 p
20. Zeisler R, Greenberg RR, Stone SF, Sullivan TM (1988) Fres Z Anal Chem 332: 612

3.3 Specimen Bank Activities at NIES

Y. Ambe and K. Okamoto

As a preliminary study for the future establishment of a full scale Environmental Specimen Bank, the National Institute for Environmental Studies (NIES), Japan, has performed an experimental research on the long term preservability of pollutants in the environmental samples such as PAH in atmospheric particulate matters and spiked chemicals in mussel meat. It has been concluded that, though not for all, some stored samples can be useful, by selecting appropriate storing method, for long term environmental monitoring.

Parallel to the above study, a small size experimental specimen bank has been operated to examine practical problems relating the management of specimen bank.

Introduction

In Japan, at present, there is no concrete program to establish a full scale Environmental Specimen Bank. While the importance and usefulness of specimen banking for long-term environmental monitoring of toxic chemicals has been recognized for several years by environmental scientists and administrative officials concerned, the realization of it is still not promising, mainly because of financial and manpower difficulties.

Some preliminary research activities, however, relating to the long-term storage of environmental samples have been performed at NIES (National Institute for Environmental Studies) cooperating with several university laboratories [1].

The activities concerning environmental specimen banking at NIES were initiated in 1980, as subprograms of a research project on the methodology of long-term monitoring of pollutants in the environment in Japan. The major part of this program, including development of highly sensitive analytical methods for inorganic and organic pollutants in remote areas, has already been terminated since 1986. The research activities directly concerned with sample banking problems were divided into the following two groups.

One was the study of the long-term preservability of pollutants in the stored environmental samples, and the other was on the various problems such as collection, transportation, storage, management of samples, and countermeasures for accidents which may occur during the operation of a full-scale specimen bank. To accomplish the above studies, a small-scale specimen bank has been operated experimentally.

Specimen Banking
Rossbach/Schladot/Ostapczuk (Eds.)
© Springer-Verlag Berlin Heidelberg 1992

Studies on the Stability of Samples

As basic studies on sample banking, experiments to investigate the preservability of chemicals in various kinds of environmental samples were designed. Subgroups of each lot of samples were subjected to different conditions of pretreatment and storage and have been analyzed at yearly intervals to ascertain changes during storage.

Stability of Benz(a)pyrene (B(a)p) in the Stored Atmospheric Particulate Matter Samples

An experiment to check the stability of B(a)p in atmospheric particulate matter samples was conducted using samples collected on quartz glass fiber filters with high volume air samplers on an urban road side in Tokyo.

Samples on the filters were stored in stainless steel boxes under the following four conditions: 1) 20 °C, in air, 2) 20 °C, in argon gas, 3) − 20 °C, in air and 4) − 20 °C, in argon gas. The concentration of B(a)p in the sample filters was measured after 0.5, 1, 2, and 4 years storage to investigate the changes occurring during storage.

The changes of the concentration of B(a)p in the stored samples expressed as a percentage of the initial values (100%) for the two series of samples with different concentrations gave nearly similar results, both decreasing with time of storage.

The average rate of decrease of B(a)p during the test period was higher for the samples stored at 20 °C decreasing by about 35% of the initial values. The samples stored at − 20 °C, on the other hand, reduced their B(a)p contents by 12% in both cases, indicating the significant effect of the storage temperature. The difference between air and argon in the storage boxes had no remarkable effect.

A rapid decrease of the concentration in the initial stage was noticeable for storage at 20 °C and − 20 °C. The difference between the rate of decrease during the first 6 months and that of the following period was significant, showing that in the first 6 months of storage the rate of decrease was approximately twice as high as that in the next 6 months; after one year storage the rate decreased almost to zero under the − 20 °C condition, whereas under the 20 °C condition the concentration still continued to decrease annually.

From the results obtained so far, it was concluded that, under the storage conditions (− 20 °C) applied in our storage program, the preservation of samples was not always complete, but, considering that the rate of reduction of B(a)p after 6 months was nearly zero, at least 80% of the initial amount can be expected to be preserved after more than 20 years storage and can be utilized for a retrospective analysis [2, 3].

Stability of Spiked Chemicals in Biological Samples

An experimental study was designed to check changes in the concentration of chemicals artificially spiked into homogenized mussel meat under different conditions of storage.

The tested samples were prepared by adding an acetone solution of several kinds of organochlorine pesticides, PCB, hydrocarbons, and other organic chemicals into homogenized mussel meat. The mussel meat was homogenized again and distributed into small containers. The variables examined were the container materials (e.g. glass bottle, glass ampoule, and polyethylene bottle) and the temperature of storage ($-20\,°C$, $-85\,°C$, and liquid nitrogen temperature).

After two years of storage, part of the stored samples were analyzed for the concentration of spiked chemicals. It was observed that organochlorine pesticides did not change at all under these storage conditions. Phtalates, on the other hand, showed a small concentration decrease in the plastic bottle at $-20\,°C$ and some organic chemicals such as N-dimethylnitrosoamine were reduced in concentration by ca. 40% at $-20\,°C$ for storage in every container type used. Unfortunately, the storage condition using liquid nitrogen had to be terminated due to a careless accident.

The analysis of spiked chemicals in the stored samples was conducted in cooperation with several university laboratories. The remaining parts of the samples are still stored under the same conditions for future analysis.

Stability of Certified Reference Materials

The preservability of certified environmental reference materials (pepperbush leaves, pond sediments, chlorella, human hair, mussel meat, vehicle exhaust particles, tea leaves, sargasso, and rice flour) which were prepared at NIES as references for chemical analysis were also programmed to be tested for changes in their elemental composition during storage at room temperature in order to check their long time availability as reference materials.

No significant change has been found so far in the certified concentrations of the elements in all the reference materials tested.

Thus, it was confirmed that these certified reference materials, which are widely used for the quality control in chemical analysis of environmental and biological samples, are stable and usable at least for five years after the time of preparation.

The Experimental Specimen Bank at NIES

In connection with the above studies on the preservability of samples, a small-scale experimental sample bank has been operated using storage facilities at NIES to examine problems related to specimen banking such as collection and transportation of samples, registration and management system of samples, maintenance of storage facilities, countermeasures against accidents (e.g. earthquakes), which may occur in the practical operation of the bank in future.

The storage facilities used in this program consist of $20\,°C$ and $-20\,°C$ storage rooms, $-85\,°C$ and $-115\,°C$ deep freezers. The storage conditions applied in this sample bank are, although not always perfect, readily available and practically most suitable, for the presently selected sample types and the capacity of the storage rooms and deep freezers.

Among the stored samples, atmospheric particulate matter samples from a remote island, lake waters of several polluted and unpolluted lakes, mussels and other marine organisms were collected under a systematic long-term monitoring program.

Samples, which were collected rather arbitrary during various environmental research works or other monitoring activities, have also been stored depending on the capacity of the storing facilities; these samples will be valuable and useful for retrospective analysis or for the study of environmental problems in the future.

Table 1 shows the list of major groups of samples stored in the bank at present. Water, sediments, soil, atmospheric particles samples and most of the biological samples are stored in a $-20\,^\circ$C storing room; some of the more important biological samples are preserved in deep freezers. Only human hair samples are stored in the bank because of ethical reasons and legal regulations in Japan regarding the collection of human tissue.

Table 1. Environmental samples stored in the NIES experimental specimen bank (as of Aug. 1990)

	Approximate number of samples
Atmospheric samples	
Atmospheric particulate matter samples collected on filters	200
Rain water & snow	50
Water samples	
Lake water	100
Sediment and soil samples	
Lake sediment	100
Soil	50
Biological samples	2000
Plant leaves	
Moss and lichens	
Fish	
Shellfish	
Birds	
Human hair	
etc.	
Certified reference materials	
Pepperbush	
Pond sediment	
Mussel meat	
Rice flour	
Human hair	
Tea leaves	
Vehicle exhaust particles	
Chlorella	
Sargasso	

The number of stored samples is increasing yearly for the limited capacity of the storing facilities, and the stored samples are programmed to be replaced by new, more valuable ones when such samples are available, although aged samples tend to be more precious with time.

This tentative specimen bank has been working without any serious problem and is planned to continue for as long as possible in order to demonstrate the usefulness of specimen banking.

References

1. Ambe Y (1984) The state of the art of researches on environmental specimen banking in Japan. In: Lewis RA, Stein N, Lewis CW (eds) Environmental specimen banking and monitoring as related to banking. Martinus Nijhoff, The Hague, p 33
2. Ambe Y (1985) An overview of the research activities relating to environmental specimen banking in Japan. In: Wise SA, Zeisler R (eds) International review of environmental specimen banking. NBS special Publication 706, p 22
3. Ambe Y, Mukai H, Okamoto K (1988) Banking of atmospheric particulate matter samples for long term monitoring of atmospheric pollution and related reference material at the National Institute for Environmental Studies. In: Wise SA, Zeisler R, Goldstein GM (eds) Progress in environmental specimen banking. NBS Special Publication 740, p 108

4 Practical Specimen Banking

4.1 Selection and Preparation of Relevant Reference Materials for Agricultural Purposes

M. Ihnat

Soil fertility, contamination and erosion influence the nutritional and toxicological qualities of agricultural food crops and the associated soils. For long-term monitoring and other studies of chemical characteristics of current and environmental specimen bank materials, incorporation of appropriate reference materials (RM's) into the analytical work is imperative for data quality control.

A summary is presented of factors to be considered in the development of biological reference materials. Some guidelines are offered regarding approaches to the generation of the varied materials required for analytical quality control. Although a moderate range of crop-based and some soil-based control materials are available for elemental analysis in agricultural science, quality control gaps are still evident, especially in the area of food materials reflecting the range of natural matrices, some important nutritional and toxic elements, and soil RM's certified for bioavailable or extractable elements. Experimental work is presented regarding the development of new crop and animal tissue materials to augment the current collection of agriculture/food reference materials. In an effort to bring to the analytical community new natural matrix RM's, ten candidate RM's, representing an extremely wide range of matrix and elemental composition, were prepared and characterized for homogeneity and recommended concentrations. An extensive interlaboratory cooperative certification campaign was completed, enabling these materials to acquire reference material status thus contributing an important additional range of natural matrices and elemental concentrations to the world repertoire of agricultural reference materials.

1 Introduction

Although biological materials in environmental specimen banks (ESB) may be used to prepare or serve as reference materials (RM's), their main and important purpose is to provide a stable, viable, collection of natural, environmental materials for retrospective chemical analyses. Such analytical determinations must be of high reliability for valid intercomparisons of data over time and between laboratories. Reference materials produced internally within the specimen bank program or externally are vital for the establishment and maintenance of analytical data quality. ESB materials in storage such as plant, and wildlife tissues, sediments and soils are related or identical to agricultural materials and thus a discussion of reference materials for agricultural purposes has a bearing on ESB activities.

A cost-effective approach to monitoring and maintaining reliability of analytical procedures is the incorporation of appropriate, compositionally-similar reference materials into the scheme of analysis. Agricultural and food commodities represent an extremely wide range of composition, in respect of the

Specimen Banking
Rossbach/Schladot/Ostapczuk (Eds.)
© Springer-Verlag Berlin Heidelberg 1992

sought-for analyte and the supporting material (matrix), not fully reflected in currently available biological RM's. This chapter summarizes the RM concept for quality control, criteria for the development of RM's available agriculturally relevant materials, and their selection, preparation, and characterization. Even though RM-producing activity has flourished in recent years, there is still a widespread deficiency in the world repertoire of suitable materials with regard to natural matrices, some important nutritional and toxic analytes, and bio-available indices. With the view to filling some of these gaps with suitable products, experimental work is summarized dealing with the selection, pre-paration and characterization of 10 candidate agricultural/food RM's for inorganic constituents.

2 Quality Control of Analytical Data and the Reference Material Concept

Analytical information on a wide range of inorganic and organic constituents in a vast array of materials provides data for research, regulatory and legal purposes in disciplines such as agricultural/food science, clinical chemistry, medical science, nutritional, and environmental sciences. Sound data are neces-sary for proper conclusions and decisions not only in regulatory and legal affairs but also in research activities. An example of the latter is the overwhelming requirement for inter-comparability of analytical data over space (between laboratories and locations) and time (temporal variation) in long-term monit-oring of soil and agricultural commodities, and in environmental specimen banking activities.

In agriculture and food sciences, accurate data on the chemical composition of raw agricultural products and foods are needed to assess effects of farm management practices and food processing on nutrient and toxic chemical content of retail food products, to establish the essentiality of nutrients or toxicology of toxicants, to identify adequate, subadequate, or marginal intakes by the population, to establish nutrient dietary requirements, to accumulate baseline concentration data, and to comply with legal labeling requirements.

The technical literature, however, abounds with examples of inconsistent information, not only in the case of trace or ultratrace analyses with high operational and competence demands, but also in the determination of major elements [1–5]. Reasons for the general lack of agreement among laboratories of the outcome of analytical work stem from the multitude of factors influencing the reliability of the final results. These factors include presampling, sampling, sample handling and manipulation, measurement, data handling and interpreta-tion, contamination control, data quality control and analyst competence.

A reference material is considered to be any material, device or physical system for which definitive numerical values can be associated with specific properties and that is used to calibrate a measurement process. Within the laboratory quality control program, incorporation of appropriate, compositionally-similar reference materials is one valuable, cost-effective aspect of a good quality control program, and a way of transferring accuracy from well defined methods of analysis to the laboratory [6–10]. Results obtained with the reference material taken concurrently through the analysis with actual samples are compared with the certified values. Closeness of agreement indicate performance of the analytical method and may suggest the need for modifications to reduce errors.

3 Criteria for the Development of Reference Materials

The development of RMs (defined as the composite of all activities leading to the final characterized product) takes into account a number of requirements and criteria [11]. These considerations which also reflect the sequence of steps to be taken in RM development are: reference material development philosophy, end use requirements, selection of materials, preparation, physical and chemical characterization, certification, documentation, and distribution.

RM philosophy includes an in-depth understanding of and appreciation for all of the concepts behind good analytical measurement, quality control, and the role of RM's. End use requirements dictate an assessment of specific details such as the nature of the problem to be addressed by the product, market requirements, analytical methods to be served, analytes, forms and concentrations to be certified, and the level of certification required. Selection of materials is made in respect of: consideration of the nature and physical form of the material in relation to real world products; representativeness of the material of the range of products in production, commerce and undergoing analysis, in respect of matrix and endogeneous and adventitious levels and speciation of analytes; the need for uncontaminated, endogenous-level natural materials; and the need for "blank" control materials and the availability of similar reference materials.

Preparation refers to all of the physical steps necessary to bring the starting material to candidate material status. It includes acquisition of commercially available, or custom prepared starting material and steps dealing with collection, cleaning, component separation, comminution, drying, sieving, blending, and packaging. Selection, testing, and assessment of preparative procedures, control and monitoring of contamination, assessment of stability expected during long-term storage, and steps for stability enhancement are other details in the preparative effort.

Both physical and chemical characterization are required to establish material properties. Physical characterization includes sieving, visual and microscopic examination, assessment of particle size distribution, moisture loss/pickup, and moisture determination techniques. Chemical characterization deals with the establishment of analyte concentration values, following an appropriate protocol, to establish material homogeneity and to provide data for recommendation/certification. Certification implies the assignment of a reliable, unassailable numerical value to a property of a material. It deals with the assessment of analytical information, selection of data and computation, and assignment of certified (recommended) values as well as associated uncertainties or confidence limits.

Adequate documentation is required in respect of the numerical values assigned to the product, instructions on its use, and information on its production and characterization. A distribution/marketing strategy is needed with the related infrastructure of storage, pricing, marketing, handling, and shipping; this is most effectively managed by a national or international standards agency.

4 Available Agriculturally Relevant Reference Materials

The discipline of agriculture cuts a broad swath through the range of existing natural materials gathering into its realm a wide variety of biological and environmental (nonbiological) materials of primary and secondary relevance. Based on the products listed by Cortes Toro et al. [12] in their most useful compilation of available reference materials, and including fertilizers, materials of agricultural relevance (agronomy, nutrition, environment, research, regulation etc.) can be grouped as follows: biological: animal body fluids and products, animal tissues, flours and cereal products, terrestrial plants, other foodstuffs, marine animals, and aquatic plants; environmental: soils, fertilizers, sediments, sewage sludges, minerals, rocks, and fresh waters.

Table 1. Suppliers of agriculturally-relevant biological and nonbiological reference materials for chemical composition quality control

Code	Name[a]
AMM	Academy of Mining and Metallurgy, Poland
ARC	Agricultural Research Centre, Finland
BCR	Community Bureau of Reference (BCR), Belgium
BOWEN	Dr. HJM Bowen, United Kingdom
CANMET	Canada Centre for Mineral and Energy Technology, Canada
CZIM	Nuclear Research Institute, Czechoslovakia
EPA	U.S. Environmental Protection Agency, USA
FISHER	Fisher Scientific Company, USA
GHENT	Department of Internal Medicine, Belgium
IAEA	International Atomic Energy Agency, Austria
ICHT	Commission of Trace Analysis of the Committee for Analytical Chemistry of the Polish Academy of Sciences, Poland
IGGE	Inst. of Geophysical and Geochemical Exploration, China
IRANT	Institute of Radioecology and Applied Nuclear Techniques, Czechoslovakia
KL	Kaulson Laboratories Inc., USA
NIES	National Institute for Environmental Studies, Japan
NIST	National Institute of Standards and Technology, USA
NRCC	National Research Council Canada, Canada
NYCO	Nycomed Pharma, Diagnostica, Norway
SHINR	Shanghai Institute of Nuclear Research, China

[a] Adapted from [12] which includes complete names and addresses

Over the past two decades, proliferation of RMs has reached the point where compilations of available products are useful [12–14]. Table 1 lists names of major suppliers of a variety of agriculturally-relevant biological and non-biological RM's; complete addresses are available in the report of Cortes Toro et al. [12]. Tables 2 and 3, respectively, summarize currently available biological

Table 2. Available biological agriculturally-relevant reference materials for inorganic chemical composition[a]

Material	Code	Material	Code
Animal Tissues and fluids			
Animal blood	IAEA-A-13	Serum	NYCO-105
Blood	KL-100-H	"	NYCO-212
Blood	KL-100-L	Urine	KL-110-H
Blood	KL-100-M	"	Kl-110-L
Bovine blood	BCR-CRM-194	"	KL-140-M
Bovine blood	BCR-CRM-195	"	KL-140-I
Bovine blood	BCR-CRM-196	"	KL-140-II
Human serum	GHENT-SERUM	"	KL-142-I
Lead in blood	NIST-SRM-955a	"	KL-142-II
	b, c, d	"	NYCO-108
Serum	KL-146-I	"	NIST-SRM-2670
"	KL-146-II		
"	KL-147-I	Whole blood	NYCO-904
		"	NYCO-905
"	KL-147-II		
"	KL-148-I	"	NYCO-906
"	KL-148-II	Animal bone	IAEA-H-5
Animal Muscle (pork)	ARC/CL-AM	Copepoda	IAEA-MA-A-1/TM
Bovine liver	CZIM-LIVER	Dogfish liver	NRCC-DOLT-1
"	NIST-SRM-1577a	Dogish muscle	NRCC-DORM-1
"	BCR-CRM-184	Fish	EPA-FISH
Human Liver	SHINR-NH	Fish flesh	IAEA-MA-A-2/TM
Mussel tissue	BCR-CRM-278	Lobster hepatopancreas	NRCC-TORT-1
Pig kidney	BCR-CRM-186	Oyster tissue	NIST-SRM-1566
Albacore tuna	NIST-RM-50		
Foodstuffs			
Wholemeal flour	BCR-CRM-189	Rice Flour	NIES-CRM-10A
Brown bread	BCR-CRM-191	Rice Flour	NIES-CRM-10B
Corn Kernel	NIST-RM-8413	Rice Flour	NIES-CRM-10C
Kale	Bowen's Kale	Wheat Flour	NIST-SRM-1567a
Rye Flour	IAEA-V-8	Rice Flour	NIST-SRM-1568
Tea Leaves	NIES-CRM-7	Brewers Yeast	NIST-SRM-1569
Potato powder	ARC/CL-PP	Mixed diet	NIST-RM-8431a
Wheat Flour	ARC/CL-WF	Total diet	ARC/CL-TD
Plants			
Aquatic Plant	BCR-CRM-060	Citrus Leaves	NIST-SRM-1572
Aquatic Plant	BCR-CRM-061	Pine Needles	NIST-SRM-1575
Olive Leaves	BCR-CRM-062	Corn Stalk	NIST-RM-8412
Single cell protein	BCR-CRM-273	Cotton Cellulose	IAEA-V-9
Single cell protein	BCR-CRM-274	Hay Powder	IAEA-V-10
Sea Lettuce	BCR-CRM-279	Pepperbush	NIES-CRM-1
Chlorella	NIES-CRM-3	Sargasso	NIES-CRM-9

[a] A selection of materials adapted from [6, 12, 13, 28]

Table 3. Available nonbiological agriculturally-relevant reference materials for inorganic chemical composition[a]

Material	Code	Material	Code
Soils			
Soil	AMM-SO-1	Soil (chesnut)	IGGE-GSS-2
Soil (calcareous loam)	BCR-CRM-141	Soil (yellow-brown)	IGGE-GSS-3
Soil (light sandy)	BCR-CRM-142	Soil (limy-yellow)	IGGE-GSS-4
Soil (amended with sewage sludge)	BCR-CRM-143	Soil (yellow-red)	IGGE-GSS-5
Soil (Ferro-Humic Podzol)	CANMET-SO-2	Soil (yellow-red)	IGGE-GSS-6
Soil (Gray Brown Luvisol)	CANMET-SO-3	Soil (laterite)	IGGE-GSS-7
Soil (Black Chernozemic)	CANMET-SO-4	Soil (loess)	IGGE-GSS-8
Soil	IAEA-SOIL-7	Soil	NIST-RM-8406
Soil (dark brown podzolitic)	IGGE-GSS-1	Soil	NIST-RM-8407
		Soil	NIST-RM-8408
Sediments			
Estuarine Sediment	BCR-CRM-277	Pond Sediment	NIES-CRM-02
Estuarine Sediment	NIST-SRM-1646	River Sediment	BCR-CRM-320
Lake Sediment	BCR-CRM-280	River Sediment	NIST-SRM-2704
Lake Sediment	CANMET-LKSD-1	Stream Sediment	IGGE-GSD-9
Lake Sediment	CANMET-LKSD-2	Stream Sediment	IGGE-GSD-10
Lake Sediment	CANMET-LKSD-3	Stream Sediment	IGGE-GSD-11
Lake Sediment	CANMET-LKSD-4	Stream Sediment	IGGE-GSD-12
Lake Sediment	IAEA-SL-1	Stream Sediment	CANMET-STSD-1
Marine Sediment	NRCC-BCSS-1	Stream Sediment	CANMET-STSD-2
Marine Sediment	NRCC-MESS-1	Stream Sediment	CANMET-STSD-3
Marine Sediment	NRCC-PACS-1	Stream Sediment	CANMET-STSD-4

Material	Reference code	Material	Reference code
Sewage sludges			
Incinerated Sludge	FISHER-SRS012	Sewage Sludge	BCR-CRM-144
Municipal Sludge	EPA-SLUDGE	Sewage Sludge	BCR-CRM-145
		Sewage Sludge	BCR-CRM-146
Rocks and minerals			
Iron formation	CANMET-FER-1	Brick clay	NIST-SRM-679
Iron formation	CANMET-FER-2	Argillaceous limestone	NIST-SRM-1c
Iron formation	CANMET-FER-3	Dolomitic limestone	NIST-SRM-88b
Iron formation	CANMET-FER-4	Potash feldspar	NIST-SRM-70a
Syenite	CANMET-SY-2	Soda feldspar	NIST-SRM-99a
Syenite	CANMET-SY-3	Glass sand	NIST-SRM-81a
Augite-olivine gabbro	CANMET-MRG-1	Glass sand	NIST-SRM-165a
Apatite Concentrate	ICJTJ-CTA-AC-1	Glass sand	NIST-SRM-1413
Feldspar	IAEA-F-1	Obsidian rock	NIST-SRM-278
Flint clay	NIST-SRM-97b	Basalt rock	NIST-SRM-688
Plastic clay	NIST-SRM-98b		
Water			
Estuarine Water	NRCC-SLEW-1	Water	NIST-SRM-1642b
Waste Water	FISHER-SRS002	Water	NIST-SRM-1643b
Water	NIST-SRM-1641b		

a A selection of materials taken from [6, 12, 13, 28, 39, 40] and nonbiological reference materials of relevance to agriculture, certified for inorganic constituents. The small collection of materials with data for organic micronutrients and radionuclides is listed by Cortes Toro et al. [12].

5 Selection and Preparation of Agricultural Biological Reference Materials

Raw and processed agricultural products, foods and feedstuffs in worldwide use and commerce constitute an extremely broad range of chemical composition with respect to major and minor elements, as well as nutrient and toxic organic compounds, ash, silica, fiber, protein, carbohydrate, and fat. The concentrations of nutritionally essential major and minor elements and toxic elements occurring therein present extremely wide ranges [15]. If one accepts the premise that for most appropriate control of analytical data quality, there should be a reasonably close correspondence, with respect to analyte and matrix composition, between reference materials and test samples, a variety of RM's must be available.

A survey of RM availability and suitability is a first step prior to embarking on the development of a specific product. Lack of correspondence between available RMs and commodities analyzed either on the basis of matrix [16, 17] or analyte composition [16, 18] suggests the need for additional appropriate control materials. Compilations of information as presented in Tables 4 and 5 serve to illustrate the current status of available RMs. From Table 4, an updated version of that in [11] one observes that although quite a number of biological RM's are available for some elements such as Ca, Cd, Cu, Fe, Hg, K, Mg, Mn, Na, Ni, Pb, Se, and Zn, there is a shortage for B, Ba, Be, Br, Cs, F, I, La, Li, N, S, Sn, and V including the agriculturally-important elements B, I, N and S. Table 5, presenting a comparison of elemental concentration ranges in foodstuffs and available biological RM's illustrates the shortfall of appropriate RM's in respect of analyte composition. The last column in the Table 5, indicating lack of coverage by existing RM's of the low and high ends of the elemental concentration ranges of laboratory test materials, shows a deficiency of available RM's in almost every instance at the low end but also frequently at the high end of the concentration range. Such information together with the desire to complement natural matrices represented by the commodity classes, cereal products, dairy products, eggs and egg products, meat and meat products, fish and marine products, vegetables and vegetable products, fruit and fruit products, fats and oils, nuts and nut products, sugar and sugar products, beverages, spices and condiments, plants and plant products, cacao bean and products, total diets, and animal feeds, was instructive in guiding the preparation and characterization of additional candidate materials. A summary of the author's experimental work carried out in the preparation and characterization of agriculture/food materials is included in the following sections.

Table 4. Ranges of certified and recommended total elemental concentration values (mg/kg or mg/L) reported for biological reference materials[a]

Element	Range[b]	No. of RM's[c]	Element	Range	No. of RM's
Ag	0.04–0.89	5 (5)	Li	–	0 (1)
Al	0.020–775	9 (4)	Mg	15–6500	36 (27)
As	0.00485–115	23 (14)	Mn	0.0077–3771	43 (29)
B	49.0	1 (1)	Mo	0.0075–3.5	18 (8)
Ba	6–79	4 (5)	N	42790–58800	2 (4)
Be	0.025–5	4 (NR)	Na	6.0–36700	27 (25)
Br	0.38–48.8	7 (8)	Ni	0.03–10	23 (13)
Ca	42–212000	35 (32)	P	592–10200	18 (14)
Cd	0.0005–26.3	46 (27)	Pb	0.017–64.4	52 (31)
Cl	550–55800	12 (13)	Rb	0.48–53.4	16 (14)
Co	0.0036–0.87	16 (12)	S	1650–15560	7 (6)
Cr	0.00076–10	23 (17)	Sb	0.005–0.07	2 (2)
Cs	0.010–0.0763	2 (2)	Sc	0.014	1 (1)
Cu	0.13–720	53 (37)	Se	0.004–20	35 (20)
F	5.87	1 (3)	Sn	0.139	1 (1)
Fe	1–2400	45 (32)	Sr	0.138–1000	12 (14)
Hg	0.0003–2.52	32 (27)	Th	0.037	1 (2)
I	0.087–5.35	5 (4)	U	0.00071–0.116	3 (4)
K	680–61000	28 (27)	V	0.386–2.3	4 (4)
La	0.0864	1 (1)	Zn	1.2–852	48 (30)

[a] Adapted from data in [6, 11–13, 16]

[b] Only concentration values denoted C by Cortes Toro et al. [12] were selected

[c] Numbers in parentheses refer to available RM's indicated in first version of this Table in 1988 [11, 16]. NR: Not reported. The following elements, included in the report by Cortes Toro et al. [12], are not reflected in certified RM's: Au, C, Ce, Er, Ga, Hf, In, Lu, Pt, Ru, Si, Sm, Te, Ti, Tl, W

Table 5. Comparison of concentration ranges of some major and trace essential and toxic elemental constituents in foodstuffs and available biological reference materials[a]

Element	Range of concentrations (mg/kg or mg/L)		RM Deficiency at low (L) or High (H) end
	in foodstuffs[b]	in reference materials[c]	
Al	0.02–1300	0.020–775	H
As	0.000–60	0.00485–115	–
B	0.002–125	49.0	L, H
Ca	0–60000	42–212000	L
Cd	0.001–5	0.0005–26.3	–
Co	0.0001–5	0.0036–0.87	L
Cr	0.002–10	0.00076–10	–
Cu	0.002–140	0.13–720	L
Fe	0.09–1200	1–2400	L
Hg	0.0006–2	0.0003–2.52	–
K	0–47000	680–61000	L
Mg	0.3–6900	15–6500	L
Mn	0.005–900	0.0077–3771	–
Mo	0.00002–6	0.0075–3.5	L
Na	0–81000	6.0–36700	L, H
Ni	0.004–8	0.03–10	–
Pb	0.001–1.7	0.017–64.4	L
Se	0.0006–12	0.004–20	L
Sn	0.2–130	0.139	H
V	0.00005–6	0.386–2.3	L, H
Zn	0.02–1600	1.2–852	L, H

[a] Data taken from [6, 11–13, 16]
[b] Estimated typical concentration ranges on fresh or dry weight basis in foods representing 12 different food classes [15]
[c] From Table 4

The goal of this undertaking was to prepare and characterize agricultural/food reference materials representing some of the food and agricultural commodity classes listed above. Following preliminary examination of a large number of potential products and consideration of dry powder flow characteristics, homogeneity, processing requirements, purity/cleanliness, and expected long-term stability, additional quantities of a limited number of materials were acquired. Processing of these led to the production of the ten candidate agricultural/food reference materials listed in Table 6 [17, 19].

All materials with the exception of bovine muscle powder were dry, powdered commercial products; the muscle powder was prepared in-house from fresh starting material [19]. They were subjected to 2.0 MRad of cobalt 60 irradiation before being brought into a moderately clean preparation room for all subsequent processing described in detail in previous publications [17, 19].

Large scale preparatory sieving was carried out using sieves constructed from polyethylene containers fitted with nylon sieve cloths with openings ranging typically from 50 to 200 μm. The requirement that the final product be composed of particles within a reasonably narrow size range to minimize consequences of possible chemical and physical dissimilarity among particles, as

Table 6. Candidate agricultural/food reference materials in preparation and elements for which recommended concentration values are anticipated from the certification campaign

Code	Material name		Code	Material name
136	Bovine muscle powder		165	Hard Red spring wheat flour
188	Whole egg powder		166	Soft winter wheat flour
186	Corn bran		184	Wheat gluten
189	Microcrystalline cellulose		183	Whole milk powder
187	Durum wheat flour		162	Corn starch

Elements

Al	As	B	Ba	Ca	Cd	Cl	Co	Cr	Cu	F	Fe	Hg	I
K	Mg	Mn	Mo	N	Na	Ni	P	Pb	S	Se	V	Zn	

well as practical aspects of preparation and yield led to selection of preparative sieving and milling conditions. Sieves were selected to yield defined fractions of macroscopically and microscopically homogeneous material. Additional size reduction of particles prior to sieving, was effected by ball milling with plastic equipment consisting of Teflon-TFE balls in a Teflon-PFA (perfluoralkyoxy-polymer) screw cap jar.

Blending was carried out in a custom-designed, 420L, poly(methyl methacrylate) blender constructed according to generally commercially accepted specifications. In order not to exceed the blender's working capacity, materials were split into two equal portions each of which was blended separately, followed by further subdivision of each of these into equal portions and blending of combinations.

Materials were packaged into 60 mL or 120 mL clear glass bottles with pulp/Saran or triseal (polyethylene)-lined white or black polyproplyene caps. All filling and packaging was done manually by weight using polyethylene scoops with target weights 25–75 g. Random units of each material were removed during packaging for physical and chemical characterization. Hermetic sealing under nitrogen or air in aluminium–nylon pouches was done for materials deemed to warrant this additional step for enhanced long-term stability. The results of this effort yielded close to 80 kg of each material typically in 1600 individual units.

6 Characterization

6.1 Physical

Physical characterization involves visual and microscopic examination, observation of appearance, color and powder flow characteristics, determination of particle size, shape and distribution by sieving and microscopy, effects of particle size on chemical composition, moisture loss/pickup, and demixing/settling during storage and transport. Particle size distribution and influences on chemical composition of two previous materials, maize leaf and kernel, have been described earlier [20].

For the current lot of materials, observations were made of the physical appearance, flow characteristics, color and particle size distribution. Test sieving was carried out on portions of starting materials to determine particle size distribution for decisions regarding preparatory sieve sizes and the need for additional particle size reduction. Test sieving gave eight fractions defined by sieve opening sizes: ∞ –425, 425–300, 300–253, 253–202, 202–150, 150–90, 90–53 and 53–20 μm; particle size distributions for the materials are presented in [17]. Macroscopic (visual) and microscopic examinations were made leading to observations of appearance to the eye described as homogeneous or heterogeneous, flow characteristic defined as good (free flowing), poor (caking) or intermediate, and microscopic appearance of the dominant and minor particles with a mention of the presence of particles deemed extraneous to the material.

6.2 Homogeneity

Homogeneity refers to the variation of the concentration of an element among test portions taken from the same and different containers established by measurements with precise techniques. Establishment of material homogeneity is the first chemical characterization to be conducted prior to the certification exercise or in concert with it if good uniformity is generally expected.

The choice of analytical technique used to test material homogeneity is wide as long as the procedure is sufficiently precise. Should, however, the experimental design allow for the simultaneous generation of analytical information for both homogeneity and certification then the procedure must live up to the rigid criteria set out for reference or definitive methods.

For the current RM's, solid sampling graphite furnace atomic absorption spectrometry (SSGFAAS) was applied to the determination of copper and lead in submilligram subsamples, and a variety of other analytical techniques via an interlaboratory cooperative effort was used to determine many other elements in subsample sizes of 100–2000 mg [21]. SSGFAAS investigations were conducted with an SM 20 Zeeman atomic absorption spectrometer designed especially for solid sampling using 0.5 mg sample weights [22, 23]. The high sensitivity of this technique, and thus the small sample sizes required provide for a critical determination of material homogeneity. Conversely, the need for small sample weights restricts the number of trace elements measurable due to low signals. In that study, Pb was measured in bovine muscle powder 136 and Cu in bovine muscle powder 136, wheat gluten 184, corn bran 186, Durum wheat flour 187, and whole egg powder 188 (the numbers refer to the codes used in Table 6). With the exception of Pb in bovine muscle powder, excellent homogeneity was found.

The analytical characterization campaign involving interlaboratory cooperative chemical analyses was designed to provide homogeneity and certification information. Preliminary homogeneity estimates, calculated as typical or median within-laboratory coefficients of variation (CV) for nine elements, Ca,

Cu, Hg, K, Mn, Na, P, Se, and Zn in six materials were presented as examples of material uniformity with respect to some typical major, minor, trace, and ultratrace elements. Upper limit homogeneity estimates ranged from 1 to 12%, typically 2% indicating good homogeneity of all six materials for these elements, ranging in concentration from 0.0004 mg/kg for Hg to 14000 mg/kg for K.

6.3 Certification

The most demanding component of the entire undertaking, that of chemical characterization for quantification or certification purposes, encompasses analyte selection based on nutritional, toxicological and environmental significance as well as availability of suitable analytical methodologies and analysts. It includes selection of certification protocols based on definitive, reference, and validated methodologies; selection of expert analysts applying conceptually different approaches; selection, development, assessment and validation of methodologies; and adaptation of statistical protocols for data analysis.

Of the three principal approaches used for certification/quantification by various ongoing RM programs: (1) in-house use of definitive and reference methods with supplementary input from cooperating outside expert laboratories, (2) cooperating laboratories – selected "expert" laboratories working independently and using different methods, (3) volunteer cooperating laboratories – round robin, the second approach was followed in this work. This alternative is viable as long as the selection process is geared to the choice of analytical chemists with the requisite expertise, with proven track records of performance, using definitive, reference, or validated methods of analysis.

In-depth analytical chemical characterization has been completed to establish recommended values for concentrations of a large number of elements. Elements of interest cover the major, minor and trace elements of nutritional, agricultural, environmental, clinical, and toxicological pertinence listed in Table 6. Cooperating laboratories applied a wide variety of analytical methods centered on atomic absorption spectrometry, flame atomic emission spectrometry, inductively coupled plasma atomic emission spectrometry, mass spectrometry, neutron activation analysis, X-ray emission spectrometry, light absorption spectrometry (molecular UV–VIS), fluorometry, electrochemistry, Kjeldahl method for nitrogen (volumetry and spectrometry), combustion-elemental analysis, volumetry (titrimetry), ion chromatography and gravimetry. These were coupled with various sample decomposition procedures as required to yield reliable concentration estimates.

7 Roles of Reference Materials

These materials represent a fair number of the principal classes of food and agricultural commodities. In terms of matrix they in fact reflect six of the important classes of food and agricultural commodities listed in Sect 5, namely cereal products, dairy products, eggs and egg products, meat and meat products, vegetables and vegetable products, and plants and plant products. They do not generally overlap with currently available RM's and thus present a potential source of different biological materials. The candidate products offer a wide range of matrix composition with ash content ranging from < 0.1%, to ca. 6%, protein nil to 80%, fat 0.1 to 45%, carbohydrate 10 to 100%, and fibre nil to 100%. These ranges should be quite instructive in evaluating effects of matrix on analytical method performance.

It is anticipated that these materials will fulfill a number of roles. Most importantly, it is expected that all will reach Biological Reference Material status. Several materials such as cellulose and starch with expected very low concentrations of inorganic analytes can very well serve as Blank Reference Materials for many routine purposes. Incorporation of such "blank" materials into the procedure will provide better measures of sample ashing or digestion blank values than use of reagents alone. These materials are expected to serve important functions in laboratories for quality control of analyses of agricultural products in general, including crops from soil/crop minor element and soil quality monitoring and research activities. They will be particularly important for data quality control in activities where, by necessity, several laboratories have to be involved.

With respect to characterization of total elemental concentrations, values for up to 28 elements (Table 6) will be recommended which, when combined with the different matrices, will make available a wide new range of element-matrix combinations. In particular, from an agricultural/food perspective, values will be available for the important minor elements Al, B, and I and major elements N and S, addressing the current deficiencies in RM's certified for these elements. Surprisingly, for example, only two biological RMs with certified values for N, a major element of overwhelming agricultural and nutritional significance, milk powder BCR-CRM-063 (5.88% N) and Bowen's kale (4.28% N) are listed by Cortes Toro et al. [12]. Concerted effort in this work has led to recommended N concentrations for all ten products ranging from 0.04% for microcrystalline cellulose to near 15% for wheat gluten adding substantially to the currently available RM's.

8 Future Requirements for the Development of Agricultural Reference Materials

Although substantial progress has occurred during the past 25 years in the development of RM's particularly biological RM's, there are still some shortcomings in the current world repertoire of RM's, a wider range of materials reflecting both matrix and analyte compositions is still required. The status of available food matrix RM's vis a vis food products with respect to inorganic elemental analytes has been assessed quantitatively [18]. In that 1985 study, concentrations of eight nutrient elements, Ca, Cu, Fe, K, Mg, Mn, Na, and Zn in 160 commonly-consumed foodstuffs in the USA were compared with levels in a number of food related certified BRMs available then. It was demonstrated that about 80% of the products had no corresponding BRM's insofar as inorganic composition was concerned. Specific food RMs required for inorganic and organic constituents have been listed by Stewart [24] as wheat bran, beef muscle, pork muscle, fish meat, chicken, lettuce, tomato, apple, orange, bean, pea, egg, cheese, edible oil, milk, and soft drink beverages. Some major new RM requirements forecast for the 1984–89 period have been reported to be: organic pollutants in natural matrices, nutrients and toxic substances in food and agricultural products and quantification of chemical species [25]. Work by Muntau [26], BCR [27], NIST [28, 29], IAEA [30], NRCC [31], Kumpulainen [32–34], and Ihnat [17, 19, 35, 36] has added relevant agricultural materials to the world repertoire.

Data is in demand for a wide range of elements reflecting nutritionally, toxicologically and environmentally pertinent analytes. Increasing interest is being expressed in elements at ultratrace levels (μg kg^{-1}) and also in the actual native forms (speciation) in which these elements occur. In addition a range of organic analytes such as nutrients (vitamins, fats, carbohydrates), and environmental toxicants is of interest. There are still a substantial number of elements for which completely acceptable methods of analysis do not exist. As, Cr, F, I, and Mn are listed as elements with conflicting methods of analysis and Co, Mo, Si, Sn, and V as elements for which suitable analytical methods for foods are lacking [37]. For tissue analyses, As, Co, Cr, Cs, F, I, Mn, Mo, Ni, Sb, Si, Sn, and V are mentioned as elements presenting considerable analytical problems [38]. Furthermore, the measurement of most elements at ultratrace levels is accompanied by analytical difficulties, and good work is very demanding of effort and talent. Determinations of element speciation as well as of organic analytes receiving increased interest are equally if not more demanding.

As opposed to total rock analyses and the determination of total concentrations of elements in biological samples normally practiced, in soil science resort is often made to labile, extractable or bioavailable measurements of elemental levels in soils. Ion absorption by plants is related to intensity and buffering power of soils as well as absorbing power, size, and root hair distribution on plant roots. Soil tests are used to obtain an estimate of the relative movement of an element from soil into the plant. These tests involve estimates of available concentrations to crops and are made by extraction with one or more of the common extractants currently proposed for a range of elements.

No reference materials certified for extractable elemental concentrations are available to monitor the usual procedures in soil science based on extraction. The philosophy of certification rests on the concept of independent methodology, that is the application of theoretically and experimentally different measurement techniques and procedures to generate concordant results leading to one reliable value for the property. Such values are thus method independent. Extractable concentrations are generated by specific procedures and are thus method-dependent, an idea which has to be rationalized with the fundamental method-independent concept in reference material certification work. Also, consideration of particle size has relevance; a product with relatively fine particles is needed to permit the usage of small samples (ca. 1 g) for analysis in order to conserve expensive reference material but the material must also reflect the real situation with respect of element availability (based on 2 mm particle sizes). Development of extractable level soil reference materials awaits resolution of these items.

References

1. Keppler JF, Maxfield ME, Moss WD, Tietjen G, Linch AL (1970) Amer Ind Hyg Assoc 31: 412
2. Parr RM (1977) J Radioanal Chem 39: 421
3. Kukai R, Oregioni B, Vas D (1978) Oceanol Acta 1: 391
4. Brown SS, Healy MJR, Kearns M (1981) J Clin Chem Clin Biochem 19: 395
5. Versieck J (1984) Trace Elem Med 1: 2
6. Ihnat M (1988) Biological and related reference materials for determination of elements. In: McKenzie HA, Smythe LE (eds) Quantitative analysis of biological materials. Elsevier, Amsterdam, Appendix 1, p 739
7. Ihnat M (1988) Biological reference materials for quality control. In: McKenzie HA, Smythe LE (eds) Quantitative analysis of biological materials. Elsevier, Amsterdam, chp 19, p 331
8. Cali JP, Mears TW, Michaelis RE, Reed WP, Seward RW, Stanley CL, Yolken HT, Ku HH (1975) The role of standard reference materials in measurement systems. National Bureau of Standards Monograph 148. Washington, DC
9. Huntoon RD (1975) In: Seward RW (ed) Standard reference materials and meaningful measurements. National Bureau of Standards Spec. Publ. 408, Washington, DC, p 4
10. Uriano GA, Gravatt CC (1977) CRC Crit Rev Anal Chem 6: 361
11. Ihnat M (1988) Fresenius Z Anal Chem 332: 568–572
12. Cortes Toro E, Parr RM, Clements SA (1990) Biological and environmental reference materials for trace elements, nuclides and organic microcontaminants. IAEA/RL/128 (Rev. 1), IAEA, Vienna
13. Muramatsu Y, Parr RM (1985) Survey of currently available reference materials for use in connection with the determination of trace elements in biological and environmental materials. IAEA/RL/128, IAEA, Vienna
14. Cantillo AY (1989) Standard and reference materials for marine science. NOAA Tech Memo NOS OMA 51, National Oceanic and Atmospheric Administration, Rockville, MD
15. Ihnat M (1982) Application of atomic absorption spectrometry to the analysis of foodstuffs. In: Cantle JE (ed) Atomic absorption spectrometry. Elsevier, Amsterdam, p 139
16. Ihnat M (1988) Sci Total Environ 71: 85
17. Ihnat M (1988) Fresenius Z Anal Chem 332: 539
18. Wolf WR, Ihnat M (1985) Evaluation of available certified biological reference materials for inorganic nutrient analysis. In: Wolf WR (ed) Biological reference materials: availability, uses and need for validation of nutrient measurement. Wiley, New York, p 89
19. Ihnat M (1987) Fresenius Z Anal Chem 326: 627
20. Ihnat M, Wolf WR (1985) Maize and beef muscle agricultural research materials. In: Wolf WR (ed) Biological reference materials: availability, uses and need for validation of nutrient measurement. Wiley, New York, p 141
21. Ihnat M and Stoeppler M (1990) Fresenius J Anal Chem 338: 455
22. Mohl C, Grobecker KH, Stoeppler M (1987) Fresenius Z Anal Chem 328: 413
23. Bagschik U, Quack D, Stoeppler M (1990) Fresenius J Anal Chem 338: 386
24. Stewart KK (1985) Problems in the measurement of organic nutrients in food products: an overview. In: Wolf WR (ed) Biological reference materials: availability, uses and need for validation of nutrient measurement. Wiley, New York, p 364
25. Taylor JK (1985) Standard reference materials: Handbook for SRM users. National Bureau of Standards Spec. Publ. 260-100, Gaithersburg, MD

26. Muntau H (1985) Ispra activities in the production of candidate biological reference materials. In: Wolf WR (ed) Biological reference materials: availability, uses and need for validation of nutrient measurement. Wiley, New York, p 109
27. Wagstaffe PJ, Belliardo JJ (1990) Fresenius J Anal Chem 338: 469
28. McKenzie RL (1990) NIST standard reference materials catalog 1990–1991. NIST Spec Publ 260, National Institute of Standards and Technology, Gaithersburg, MD
29. Wolf WR, Iyengar GV, Tanner JT (1990) Fresenius J Anal Chem 338: 473
30. International Atomic Energy Agency (1990) Intercomparison runs, reference materials. IAEA, Vienna
31. Marine reference materials and standards (1990) National Research Council of Canada, Marine Analytical Standards Program, Ottawa
32. Kumpulainen J, Paakki M, Tahvonen R (1988) Fresenius Z Anal Chem 332: 685
33. Kumpulainen J, Paakki M, Tahvonen R (1990) Fresenius J Anal Chem 338: 423
34. Kumpulainen J, Tahvonen R (1990) Fresenius J Anal Chem 338: 461
35. Ihnat M (1986) Report of analysis, Reference material 8412, Corn (Zea Mays) Stalk, NBS, Gaithersburg, MD
36. Ihnat M (1986) Report of analysis, Reference material 8413, Corn (Zea Mays) Kernel, NBS, Gaithersburg, MD
37. Beecher GR, Vanderslice JT (1984) Determination of nutrients in foods: factors that must be considered. In: Steward KK, Whitaker JR (eds) Modern methods of food analysis. Avi, Westport, CT, p 29
38. Smith JC Jr., Anderson RA, Ferretti R, Levender OA, Morris ER, Roginski EA, Veillon C, Wolf WR, Anderson JJB and Mertz W (1981) Fed Proc 40: 2120
39. Bowman WS (1990) Certified reference materials. CCRMP 90-1E, Canada Centre for Mineral and Energy Technology, Ottawa
40. Xie X, Yan M, Wang C, Li L, Shen H (1989) Geochemical standard reference samples GSD 9-12, GSS 1-8 and GSR 1-6. Geostand Newslett 13: 83

Contribution No. 91-60 from Land Resource Research Centre.

4.2 The Common Mussel (Mytilus edulis) as Marine Bioindicator for the Environmental Specimen Bank of the Federal Republic of Germany

J.D. Schladot, F.W. Backhaus

The mud flats as part of the North Sea represent a particularly sensitive and ecologically valuable part of the Atlantic. This area takes on a key role in the ecosystem of the North Sea and is characterized by a rhythmic alternation between high and low tide. The mud flats are a large-scale biotope for various fish species, mussels, crustaceans, gastropods, etc. At the same time the mud flats also represent a residential and economic area for the human population. Because of this reason, serious problems affect the mud flats and the North Sea coast from different interferences which are not environmentally compatible. Environmental pollution due to anthropogenic influences is still on the rise and requires comprehensive controls and monitoring strategies. Within the Environmental Specimen Bank of the Federal Republic of Germany, mussels of various representative areas of the mud flats are periodically collected and analyzed. Time dependent trends require precisely described collection, sample preparation and analytical characterization. Therefore, Standard Operation Procedures (SOPs) are prepared for the collection of mussels and other specimen. The selection of a representative collection area for mussels is based on a screening test. Some results of a screening test in the Schleswig-Holstein mud flats are given in this report. The collection of mussels for the Environmental Specimen Bank is done in the "Königshafen List" a part of the Sylt-Römö-mud flats in a two month period.

General Ecological Principles for the Significance of Marine Bioindicators in Tidal Mud Flats

The North Sea is an arm of the Atlantic of particular ecological and economic significance. It is a sensitive marine area, rich in fish and invertebrates, and is extremely productive and biologically active.

The mud flats stretching along the coast of the German Bight from Den Helder to Esbjerg represent a particularly sensitive and ecologically valuable part of the North Sea. This area takes on a key role in the ecosystem of the North Sea and is characterized by a rhythmic alternation between high tide and low tide. The habitat of the mud flats is divided into several subareas each colonized by different biocoenoses: diatoms on the mud flats are of great significance for the mud flat fauna in their capacity as oxygen producers and food; worms, mussels, gastropods, and small crustaceans live in and on the tidal mud flats; at low tide numerous fish take refuge in the tideways with permanent water flow.

The mud flats are a large-scale biotope for about 25 typical fish species from the North Sea, of which several are mainly dependent upon the mud flats at

Specimen Banking
Rossbach/Schladot/Ostapczuk (Eds.)
© Springer-Verlag Berlin Heidelberg 1992

certain stages of their development. The fish species make use of the mud flats in different ways, for example:

Several species spend their entire life on the mud flats (sedentary fish)
Other species spend their juvenile phase on the mud flats (nursery)
Still other species use this area at certain seasons (migratory fish)

In addition, the mud flats are an area of central importance for bird populations [1]. The Schleswig-Holstein mud flats are of the greatest significance as a moulting area for sheldrakes and eider ducks. More than 20 species of coastal birds dependent on special breeding grounds breed within the mud flats including such rare species as plover, common tern, and oyster catchers. Apart from the seagulls, most of the breeding birds are on the red list of endangered species.

Furthermore, numerous biotopes are present on the mud flats where about 2000 smaller animal species live, many of which reach their greatest numbers in this coastal habitat in the shallow sea.

However, the mud flats also represent a residential and economic area for the human population. Serious problems affecting the mud flats and North Sea coast result from interference with the landscape, e.g. dyke construction which is not environmentally compatible. Furthermore, modern industrialized society has also caused chemical changes. The unusual introduction of nutrients into the sea, from which they cannot escape again, causes a massive growth of algae. The North Sea is also intensively used by the bordering countries. This utilization involves pollution and endangerment via the following input paths:

Introduction via rivers, atmospheric input, discharge of municipal and in-dustrial waste water from the land, marine waste disposal, drilling platforms, shipping, and others.

The input of nutrients via the rivers due to fertilizers and via municipal wastewater increased dramatically up to 1980; since then it has remained at four times the natural river load for nitrogen and seven times the natural level for phosphorus. Excessive quantities of nutrients in the seawater may lead to a massive development of plankton. The biochemical degradation of this excess production in water layers close to the bottom leads to great oxygen consumption. Oxygen deficiency may then cause the mass death of higher animals such as fish, crustaceans and mussels.

Heavy metals cannot be degraded. They may accumulate in sediment and organisms (bioaccumulation) and may reach such concentrations that toxic effects result. Heavy metals enter the North Sea in the form of industrial waste products. Inputs via rivers, the atmosphere, and direct discharges are all of the same magnitude. The pollution of the North Sea with heavy metals has led to a change in flora and soil fauna and to a particularly disturbing increase in fish diseases [2].

Many pollutants are taken up from the seawater by the animals and deposited, for example, in the liver, muscles, brain, and fatty tissue. This therefore leads to an accumulation in the food chain and for this reason the final links in the food chain (e.g. seal, man) are most seriously affected.

Organohalogen compounds mainly enter the environment from the chemical industry, or chemical-related industries. The organohalogen compounds detected in the North Sea include the following:

1) Polychlorinated biphenyls (PCB). They are used as coolants, hydraulic liquid, transformer oil and as softeners for plastics. Furthermore, PCBs are also found in used oil.
2) Hexachlorobenzene (HCB), lindane (gamma HCH), DDT, dieldrin. HCB is used, for example, for seed dressing and as a softener for PVC. The others are pesticides.

Due to their chemical properties, these non-natural stable substances are accumulated in the fatty tissue of organisms. Organochlorine compounds are not, or only very slowly, degraded and even the lowest concentrations may have a toxic effect. In the case of fish, the accumulation of these substances in the gonads may impair the reproduction capacity. Furthermore, high concentrations of these substances have been determined in mussels, sea birds and seals, affecting their health.

During the first half of the 20th century, only a few instances of filamentous green algae occurred in sheltered bays. Since the late seventies these algae tend to proliferate in summer. They then cover wide areas of the mud flats with carpets of algae. In July 1989, for example, large areas of the East Frisian mud flats were green with algae. This proliferation of algae is probably a response to the increased nutrient contents in the coastal waters. Algae growth has also greatly increased in the plankton of the mud flats. This may possibly be a reason for the expansion of the mussel population, particularly the common mussel (*Mytilus edulis*). For example, the area of all the mussel beds in Königshafen near Sylt increased from 7 to 34 hectares from the thirties to the seventies and in Jadebusen from 170 to 430 hectares. Common mussels mainly form natural beds in the tidal regions and are often thickly covered with seaweed and represent a biocoenosis rich in species [3].

However, common mussels do not only filter plankton algae, they also rapidly release nutrient salts such as ammonium and silicate. They therefore speed up the process of plankton production. Ooze is deposited between the mussels and is then stirred up by rough seas. This causes considerable oxygen consumption in water close to the bottom. Accumulations of toxic substances in the ocean cause the mass death of mammals (e.g. seals) as well as an increase in diseases of organisms living on the mud flats, particularly the fish population in river estuaries [2]. In the final analysis, these phenomena can also harm humans via the food chain. This damage does not only refer to human health, agricultural nutrients also harm sea fishing. Greatly expanded mussel harvesting may lead to the total destruction of microecosystems if the mussel beds disappear or may indirectly result in erosion of the mud flats due to human influence. The bottom of the mud flats is not only determined by the common mussels but, above all, by the thick coating of diatoms.

Environmental pollution due to anthropogenic influences is still increasing – the number of substances and their metabolites annually entering the environment is in the order 1×10^5 and requires comprehensive controls and monitoring strategies [4]. It is therefore absolutely indispensable to go beyond the already existing legislation on limiting emissions and detecting environmental chemicals [5] and to employ additional measures for environmental monitoring since only a very small fraction of the substances entering into the environment can be measured regularly. Apart from measurements already under way of substances recognized as harmful (so-called "monitoring"), retrospective data regarding changes of concentration in indicators which can be specifically allocated in time are also important. Dated material was previously used for this purpose, on the assumption that it did not significantly change in the course of time – e.g. undisturbed sediments, glacier ice, bones, teeth, organisms from biological collections [6]. However, this involved certain problems: precise dating was frequently difficult, the material was not representative, and combination, migration, and degradation processes led to results of little significance [7].

Data reliable in every respect can therefore only be expected if representative samples are taken at certain points in time and at defined locations according to precisely described standard operation procedures (SOPs) [8] and stored in such a way as to prevent any subsequent chemical change. The storage of authentic material from various sectors of the environment should be used to determine substances which may endanger the environment. This should also enable substances not recognized as hazardous at the time of sampling or not yet capable of being analysed with sufficient precision to be determined in retrospect, to permit their trends to be followed and thus indirectly to arrive at prospective statements. The availability of reference samples capable of being stored almost indefinitely could furthermore open up new analytical possibilities.

An important aspect is the appropriate selection of a study area. Starting from a study in the literature about a core network of representative and well defined sampling areas [9, 10, 11], field work on the selection of areas was begun in early 1985. The prerequisite for the sampling area, apart from its basic suitability for sampling – sufficient population density of the most important sample species, possibility of legal protection for the sites – was, that ecological research was already being carried out or planned for the immediate future in these areas so that a concentration of important research activities resulted.

Prerequisites for the Sampling of Common Mussels

The sampling guidelines (Standard Operation Procedures, SOPs) for the German Environmental Specimen Bank for brown algae (*Fucus vesiculosus*), common mussels (*Mytilus edulis*) and herring gull (eggs) (*Larus argentatus*) currently envisage sampling every two years in two sampling areas of the mud flats. In the following, the sampling guideline (SOP) for common mussels will be discussed in detail.

Many pollutants which now reach the coastal waters in a dissolved form have such low concentrations that they are difficult to determine without an accumulation. This becomes very clear with the different accumulation of a pollutant in two different matrices (see Fig. 1). Furthermore, the accumulation methods are only designed for special pollutants. Of the many marine organisms, the edible mussel (*Mytilus edulis*) is one of those which accumulate pollutants in a dissolved or particulate form [12].

The following conditions were drawn up as criteria for the selection of common mussels (*Mytilus edulis*) as a bioindicator:
worldwide distribution, sedentary, long-lived species, sufficient size, easy to take samples, great abundance, ability to live in brackish water, and concentration of many pollutants as a filter feeder [13, 14, 15, 16].

The status of the common mussel as a specimen species for the Environmental Specimen Bank of the Federal Republic of Germany can be regarded as finally accepted.

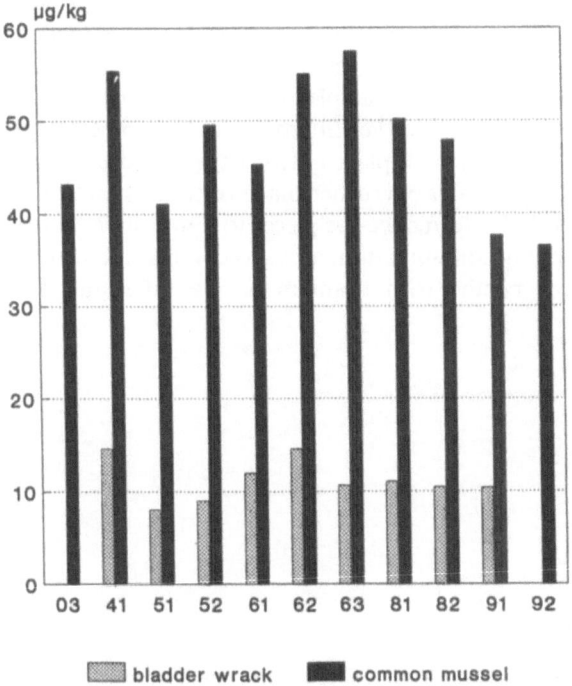

Fig. 1. Lead accumulation in two different marine organisms – brown algae and common mussel – collected at the same collection sites at the same time

Standard Operation Procedure (SOP): Sampling Plan for Common Mussels

The mussels collected during 1985–1991 are collected every two months during the sampling year at the same sampling site. About 1000 mussels ranging in size from 4–7 cm are collected during each sampling month. In addition, important environmental parameters are documented during sampling and, if necessary, a biological characterization is carried out (see SOPs) [17]. In the same way, data about the growth and population density are also important. After preliminary cleaning or removal of foreign material directly at the sampling site, i.e. more or less thorough cleaning depending on the species but always contamination controlled, the collected material is deep frozen in cryovessels approved for road transport and then transported to the Environmental Specimen Bank at temperatures of < − 150 °C [18, 19]; the material remains at this temperature during all operating steps until final storage.

Standard Operation Procedure (SOP): Sample Preparation and Storage

The samples are dissected with titanium or quartz glass knives, i.e. the soft tissue parts are removed from the shells. The soft tissue material is then directly deep frozen in liquid nitrogen. Some of the mussels are prepared for an individual analysis in order to document the variability of the pollution concentration within a mussel bed. After this the remaining sample material is processed into a homogenate under cryogenic conditions in the Environmental Specimen Bank at Jülich (see Fig. 2). Homogenization takes place in two stages: the deep frozen material first passes through a preliminary crusher made of stainless steel and cooled with liquid nitrogen in order to reduce the pieces to a maximum size of 15 × 15 mm. In order to avoid any contamination all parts coming into contact with the sample material are coated with titanium nitride. After this fine

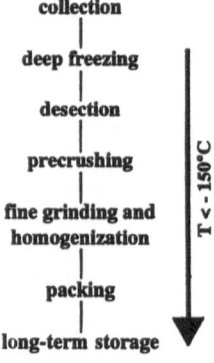

Fig. 2. Flow chart of the standardized sample preparation process for environmental specimen banking purposes under cryogenic conditions.

grinding takes place in a cryovibration mill specially developed for the requirements of the Environmental Specimen Bank [20, 21]. Depending on the sample material, i.e. its trace metal content, grinding equipment of stainless steel, titanium or PTFE can be used.

The grinding process results in a very fine powder free-flowing at low temperatures whose average grain size is less than 200 μm. An initial quantity of ca. 5 kg per specimen type and site provides 500–600 standardized subsamples. A certain number of subsamples, selected by a so-called random number generator, serve for the initial analytical characterization of the homogenate. The remaining subsamples provided with a bar code for identification purposes are stored under cryogenic conditions over liquid nitrogen (T < − 150 °C).

Implementation of a Screening Experiment with Common Mussels in the Sylt-Römö Mud Flats

A large number of potential sampling sites were studied in the detailed coverage of new study areas. So-called screening operations are carried out for these studies in the selected study area. The Schleswig-Holstein mud flats with an area of 285 000 ha represents the largest national park in Central Europe. The boundaries of the national park are formed by the German-Danish frontier in the north; a line at a distance of 150 m from the seaward edge of the dykes in the east; the northern mud flat edge of the main water way of the River Elbe in the south; and the north and east coast of the islands of Sylt and Amrum in the west. The mud flat bay between the islands of Sylt (with the Hindenburg embankment) and Römö (with the Römö embankment) at the northern tip of the Schleswig-Holstein Mud Flat National Park covers an area of approx. 400 km². 50% of this area is the main eulitoral habitat. The only connection to the North Sea is the List Creek between the islands of Sylt and Römö, 2 km wide at its narrowest point and up to 38 m below sea level. The tidal range in the Sylt-Römö mud flats amounts to 1.7–1.8 m. A channel about 2 km in length forms the only connection between the Sylt-Römö mud flats and the North Sea. It is therefore possible to undertake a balance of material flows within the Sylt-Römö mud flats without undue difficulty. The inflow of fresh water is slight in this part of the tidal mud flats.

The study area selected for the screening test – Königshafen List – is a bay 4 km in length, 1 km in width with an area of 3.8 km². It is open towards the east and is drained by a deep tideway in the centre towards the List Creek (see Fig. 3). About 88% of Königshafen is dry at low water. Approx. 55% of the area is covered for less than 6 h. A large number of different types of mud flats are represented within the area of Königshafen. The sand mud flats cover the greatest area but ooze mud flats, eelgrass, and mussel beds are also represented. Königshafen is accessible from three sides at low water. This is particularly important for implementing field missions (short transport paths, no need for boats).

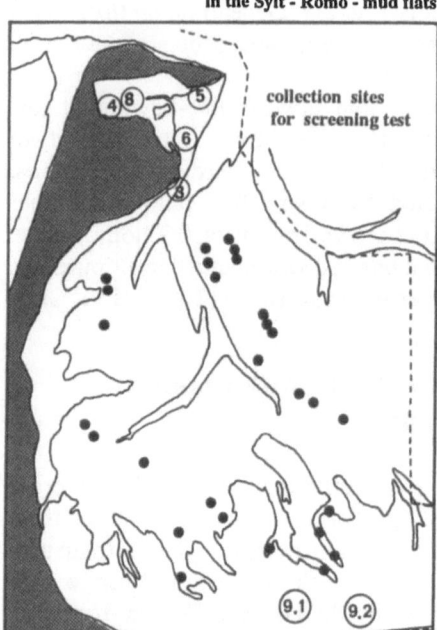

**common mussel population
in the Sylt - Römö - mud flats**

collection sites
for screening test

Fig. 3. Collection sites for the screening test in
the Sylt–Römö mud flats

The screening sites were selected on the basis of a common mussel distribution map (see Figs. 3 and 4) and an inspection of the terrain. With one exception (position 3, Fig. 3, provisional sampling site for brown algae and common mussels for the Environmental Specimen Bank studied since 1985) only those common mussel beds in the eulitoral were selected which had colonized the natural mud flat sediment. The following sampling sites were selected for the screening experiment:

4: Königshafen, western edge close to the air-to-ground firing range;
5: Königshafen, northern edge close to the List Creek (with positions 5.1 and 5.2);
6: Königshafen, southern edge close to the discharge from the List sewage works (with positions 6.1, 6.2 and 6.3);
8: Königshafen, centre (with positions 8.1 and 8.2);
9: Morsumkliff, northern Hindenburg embankment (with positions 9.1 and 9.2);
3: Sampling site provisionally studied since 1985, stone groyne close to the spa resort centre.

Two hundred individual mussels corresponding to the SOPs for common mussels [8] with a size range 4–7 cm were collected from each sampling site. The mussels were then cleaned in the water from their habitat and left to drain for

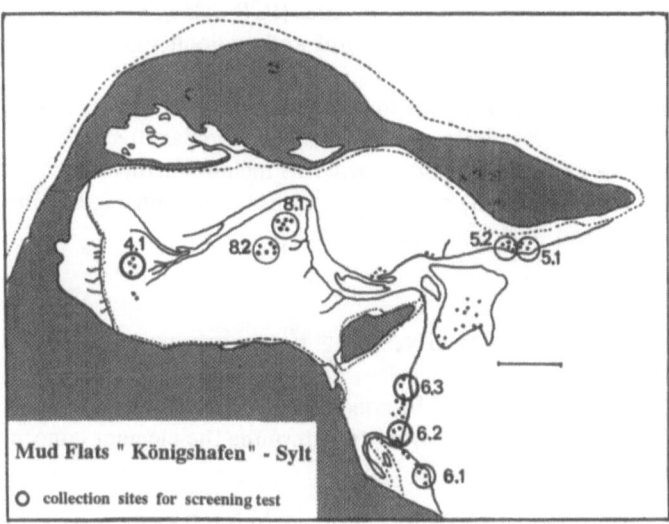

Fig. 4. Collection sites for the screening test for common mussels and brown algae in the mud flats Königshafen Sylt

about 1 h. Immediately after this the mussels were deep frozen in stainless steel vessels of 3.5 l capacity in the gas phase over liquid nitrogen. The common mussels were further processed in a clean-room laboratory at the Environmental Specimen Bank in Jülich. The flow chart in Fig. 2 shows the sample preparation process. After thawing slightly the common mussels were dissected using titanium or quartz glass knives. The soft tissue fraction is directly collected in liquid nitrogen and subsequently reduced in size in a preliminary crusher.

Standard Operation Procedure (SOP): Sample Characterization

During sample preparation the grain size distribution and the homogeneity, which is of particular significance since each subsample must be representative of the entire material, is determined by various methods such as solid AAS [22, 23] for metals. A moisture determination is finally carried out in the homogenate since the analytical data refer to both the wet and dry weight.

Modern methods are applied for the analytical characterization and are continuously adapted to the state of the art. Ten or twenty mussels per sampling site are studied for the individual characterization of the common mussels. After various types of decomposition process, adapted to the determination methods and sample species, a fingerprint is drawn up for, at present, up to 20 essential and nonessential elements and methyl mercury. The following methods are used for this purpose: all varieties of atomic absorption spectrometry, atomic emission spectrometry with inductively coupled plasma (ICP–AES), voltammetry, isotope dilution mass spectrometry, instrumental and radiochemical neutron activation analysis [24, 25].

Matrix-identical control and reference materials similarly stored in a deep frozen state [26] guarantee particularly precise values for comparative measurements. New methods are being developed in order to include further classes of compounds in the characterization, e.g. to determine surfactants and complexing agents in stored and fresh samples.

Discussion of the Results of the Screening Experiment

Only a certain selection of elements will be considered in discussing the results. The following measuring methods were used to determine the element concentrations:

Cold-vapor atomic absorption spectrometry (CV–AAS) for Hg
Hydride atomic absorption spectrometry (Hy–AAS) for As
Anodic stripping voltametry (ASV) for Pb, Cd, Ni
Graphite furnace atomic absorption spectrometry (GF–AAS) for Cu

On the basis of the inorganic investigation results for Hg, Cd, Pb, Cu, Ni, and As it can be concluded that:

1. The mussel beds at positions 4 and 9 (see Figs. 3 and 4, Sylt–Römö mud flats or Königshafen List) are unsuitable as permanent sampling sites for the Environmental Specimen Bank due to increased values for Cu, Pb, Hg, Cd. The influence of the air-to-ground firing range situated in the vicinity becomes apparent at the sampling site 4 (see values in Table 1 and Fig. 3). Furthermore, sampling in the ooze mud flats (pos. 4) can only be carried out with difficulty and at absolute low water. Increased Pb, Cd, and Ni values were measured at the sampling sites 9.1 and 9.2. Furthermore, in the additional organic analysis of polycyclic aromatic hydrocarbons, the highest PAH concentrations in common mussels were measured in this screening experiment [personal communication Prof. Jacob] at positions 9.1 and 9.2. The increased concentrations in the inorganic and also organic sector may possibly be attributed to the vicinity of the Hindenburg embankment (railway ferry). The greatly increased PAH concentrations seem to be a clear indication of this.
2. The common mussels collected at the mussel bed positions 6.1–6.3 show increased values for Pb, Cd, Cu, and Hg. This may be a result of the discharge of the List sewage works from which pollutants may enter this region of the Sylt–Römö mud flats.
3. On the basis of the inorganic analyses, the two mussel bed positions 5 and 8 in the eulitoral of the Sylt–Römö mud flats and Königshafen List thus remain for sampling in the selected area of study. Due to the geographic circumstances, position 5 should be preferred with respect to easy sampling of common mussels

Table 1. Metal content in common mussel from different collection sites of the Sylt-Römö-mud flats

Sampling site	Hg µg/kg	std. deviation	Pb mg/kg	std. deviation	Cu mg/kg	std. deviation	As mg/kg	std. deviation
3	47.9	6.5	2.9	0.6	10.61	3.7	12.6	4.27
4	60.3	10	3.45	1.1	10.2	4.3	11.85	6.15
51	38.1	3.2	1.63	0.8	8.69	1.1	16.83	2.79
52	49.52	0.9	2.22	0.7	8.07	1.3	16.39	3.16
61	52.55	17	2.65	1.1	7.72	1.1	16.84	3.45
62	82.6	15	2.22	0.9	8.3	1.1	17.2	3.22
63	67.8	12.7	2.48	0.7	6.8	1	18.23	3.55
81	53.8	4.8	2.05	0.5	7.62	1.1	13.47	4.72
82	52.4	5.8	2.24	0.3	7.38	0.5	12.89	4.39
91	42.9	10	2.78	1	6.6	0.6	15.78	4.49
92	40.9	4.7	2.11	0.6	6.9	0.9	17.11	6.7

Table 2. Mercury content in randomly selected individual common mussel and homogenate mussels from the same mussel bed

	Sampling site No. 3 Hg μg/kg	Sampling site No. 6.1 Hg μg/kg	Sampling site No. 6.2 Hg μg/kg	Sampling site No. 6.3 Hg μg/kg
Individual mussel	60.1	45.3	89.6	77.2
	45	44.7	66.2	64.8
	53	42.2	98.9	103
	44.2	78	64.5	52.8
	46.1		96.5	41
	41.2		79.9	
	45.5			
Mean	47.9	52.55	82.6	67.8
Mean of homogenated Sample (120 mussel)	43.2	45.35	55.1	57.5

even with a higher water level since this mussel bed is only a few metres from land whereas the mussel bed position 8 is in the centre of Königshafen in the immediate vicinity of the tideway with permanent water flow. Furthermore, the sampling site 5 is close to the List Creek where material balances may be initiated.

Acknowledgements: The authors would like to express sincere thanks to their colleagues in the Biogeography Department of the University of Saarland, Ch. Horras, R. Klein, M. Paulus and J. Spranger, who were responsible for selecting the sampling sites. Thanks are also due to colleagues in the Institute of Applied Physical Chemistry at the Research Centre Jülich for investigating the essential and non-essential trace elements, U. Bagschick, M. Burow, Y.J. Cho, M. Froning, G. Giernich, C. Mohl, and P. Ostapczuk, and last but not least, N. Commerscheidt and B. Süßenbach, who assisted in sampling and sample preparation.

The authors are grateful to the Federal Ministry for the Environment, Natural Protection and Reactor Safety and the Federal Environmental Agency for financial support within the framework of the Environmental Specimen Bank Project.

Particular thanks are finally also due to Dr. M. Stoeppler, who was constantly available to deal with our enquiries and whose commitment and promotion has definitively shaped the Environmental Specimen Bank in Jülich since 1975/76.

References

1. Nehls G (1989) Occurrence and food consumption of the common eider (Somateria mollisma) in the Wadden Sea of Schleswig Holstein, Helgoländer Meeresunters, 43: 385

2. Möller H (1989) Daten zur Biologie der Elbfische, Heino Möller, Kiel
3. Reise K, KOSMOS, Dez. 1990
4. Kayser D, Boehringer UR, Schmidt-Bleek F (1982) Environ Monit Assessm 1: 241
5. Gesetz zum Schutz vor gefährlichen Stoffen (Chemikaliengesetz) Bundesgesetzblatt I, 1718 (1980)
6. MARC (1985) (Monitoring and Assessment Research Centre) Historical Monitoring, Techn. Report No 31, University of London
7. Tuthill C et al. (1982) Environ Monit Assessm 1: 189
8. Umweltbundesamt Berlin, Guidelines for Environmental Specimen Banking (in press)
9. Lewis RA et al. (1985) Richtlinien für den Einsatz einer Umweltprobenbank in der Bundesrepublik Deutschland auf ökologischer Grundlage, Universität des Saarlandes
10. Lewis RA, Klein B (1990) A Brief History of Specimen Banking: Storage, Institutions and Applications, Toxicological and Biological Chemistry, Vol. 27: 251–266
11. Paulus M, Horras Ch, Klein B, Lewis RA (1990) Vertiefte Auswahl von Probenahmeregionen für die Umweltprobenbank und ökologische Beratung zu ihrem Betrieb, Universität des Saarlandes
12. Schulz-Baldes Die Miesmuschel Mytilüs edulis als indikator für die Blei-konzentration im Weserästüar ünd in der Deütschen Bücht Mar. Biol. 21, 98–102
13. National Academy of Sciences: The International Mussel Watch, 248 pp., Washington: National Academy of Sciences, 1980
14. Phelps DK, Galloway WB (1980) A report on the coastal environmental assessment stations (CEAS) program. Rapp. P. -v. Reun. Cons. perm. int. Explor. Mer. 179: 76–81
15. Ritz DA, Swain R, Elliott NG (1982) Use of the mussel Mytilus edulis planulatus L in monitoring heavy metal levels in seawater. Aust. J. mar. Freswat. Res. 33: 491–506
16. Farrington JW, Goldberg ED, Risebrough RW, Martin JH, Bowen VT (1983) U.S. "mussel watch" 1976–1978: An overview of the trace-metal, DDE, PCB, hydrocarbon and artificial radionuclide data. Envir. Sci. Technol. 17: 490–496
17. Standard Operation Procedures, in press, e.g. Oxynos K, Schmitzer J, Kettrup A, Dürbeck HW, this volume
18. Stoeppler M, Schladot JD, Dürbeck HW (1989) Umweltprobenbank in der Bundesrepublik Deutschland, Teil 1: Realisation von Umweltprobenbanken, GIT Fachz. Lab. 33: 1017–1020
19. Stoeppler M, Schladot JD, Dürbeck HW (1989) Umweltprobenbank in der Bundesrepublik Deutschland, Teil 2: Betrieb der Umweltprobenbank seit Januar 1985, GIT Fachz. Lab. 33: 1119–1124
20. Schladot JD, Backhaus F, Reuter U (1985) Beiträge zur Umweltprobenbank I: Studie zur Probenhomogenisierung bei tiefen Temperaturen unter Berücksichtigung der für die Umweltprobenbank notwendigen Parameter, Jül-Spez-330, ISSN 0343-7639, KFA Jülich
21. Schladot JD, Backhaus F (1988) Preparation of Sample Material for Environmental Specimen Banking Purposes – Milling and Homogenization at Cryogenic Temperatures, in S.A. Wise, R. Zeisler, G.M. Goldstein (eds): Progress in Environmental Specimen Banking, NBS Special Publ. No 740: 184–193
22. Mohl C, Grobecker KH, Stoeppler M (1987) Fresenius Z. Anal. Chem. 328: 413
23. Bagschik U, Quack D, Stoeppler M (1990) Fresenius J. Anal. Chem. □
24. Stoeppler M, Schladot JD, Dürbeck HW (1991) Analytiker-Taschenbuch, Band 10, Springer, Berlin Heidelberg New York (in preparation)
25. Roßbach M, Stoeppler M (1988) Fresenius Z. Anal. Chem. 332: 636
26. Roßbach M (1990) Ber. KFA Jülich, Jül-Spez. Nr. 576

4.3 The Role of Lead and Cadmium Reference Samples in an Epidemiological Case Study at Santo Amaro, Bahia, Brazil

T.M. Tavares

During a period of 24 years a 30,000 tons/year lead smelter operated in the vicinity of Santo Amaro (30,000 inhabitants). This plant used local galena as ore which contained high levels of cadmium, and no proper air pollution control system existed. About 650 children aged 1–9 years, all of low social and economic standards, lived within 900 m of this smelter. Smelter dross spread over gardens, backyards, streets, and in the home chimney filters were used as carpets, bed spreads, and rags. The children population was racially mixed; it included partially malnourished and parasitically infected (by hookworm) children. In 1980 a cross-sectional study on these children was carried out. The levels of cadmium, lead, and zinc protoporphirin (ZPP) found in the blood were among the highest in the world, leading local government to immediately enforce several costly measures to reduce the pollution level. A second cross-sectional study in the children population aged 1–9 years was conducted in 1985: Although a significant reduction in the levels of the blood indicators was observed, new cases of Pb intoxication were still occurring. The plant involved paid for treatment of the 31 children considered as under risk of intoxication. Further studies were carried out to identify the causes of the persistant exposure to the two metals. Fugitive emissions from the industry was one important cause of lead exposure. Presence of dross in homes accounted for significant exposure to cadmium. Additional antipollution measures on the part of the smelter industry are necessary.

The role of different lead and cadmium reference samples, especially those of blood, is discussed in the context of this case study.

Introduction

The town of Santo Amaro da Purificacão, with ca. 30,000 inhabitants, is situated north of the Todos os Santos bay, at the bank of the Subaé River, in the state of Bahia, Brazil (Fig. 1). During the colonial period, sugar cane production and boat transported commerce were the dominating activities in the area. Because of socio-economic changes (abolishment of slavery, construction of roads, etc), there was a reduction in these activities, and in the middle of this century some industries were installed in the area to make use of available labor – among them a primary lead smelter in 1956. The initial production of 5,870 tons Pb/year [1], with little pollution control, in an area where emissions were not dispersed to any great extent in the atmosphere, was enough to cause the death of bovine cattle. As a result, legal proceedings were issued against the lead smelter by the local farmers. The industry settled the issue by purchasing the lands closer to the industry.

During the following twenty years, the industry increased its production to 30,000 tons Pb/year with little improvement of its emission control systems, and

Specimen Banking
Rossbach/Schladot/Ostapczuk (Eds.)
© Springer-Verlag Berlin Heidelberg 1992

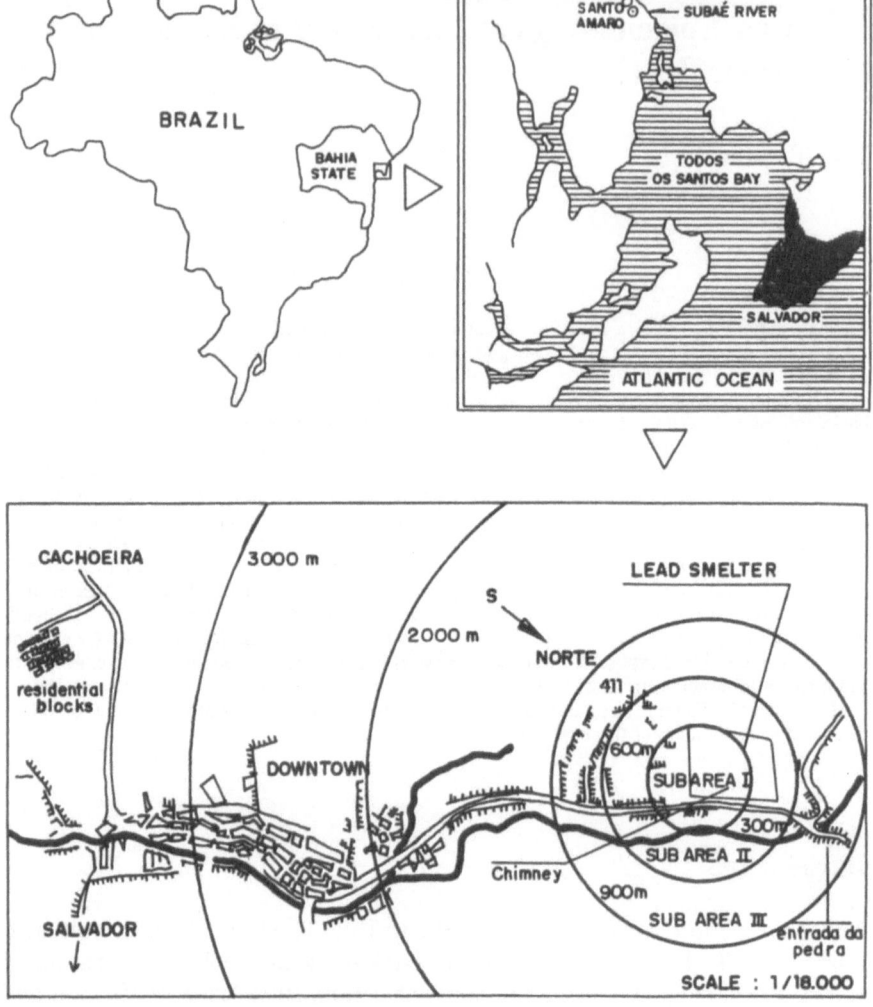

Fig. 1. Location of Santo Amaro City and its surroundings.

urban expansion resulted in new settlements closer to the smelter – particularly people of low social economic standards [2]. In 1976 the industry admitted having emitted a total of 400 tons of cadmium to the environment, 150 tons of it to the air. Even greater amounts of lead, both to the air and to the Subaé river were also emitted by this smelter. By this time, the concentration of cadmium (on a dry basis) in the local sea food was in the range 80–135 ppm for oysters, 13–40 ppm for soft shell crabs, and 40–60 ppm for local molluscs [3], but the consumption of this food did not correlate positively with the levels found in the main consumers, namely fishermen [4, 5]. The atmospheric emissions from the smelter took place through nine small chimneys, each one connected with different stages of the industrial process. The heights of these chimneys above ground level were in the range 9.5–15.5 m, but, because the industry was located in a valley, the heights relative to the surrounding elevations were lower. In 1970, primitive filters were installed in some of the chimneys. Some years later, when the old filters were exchanged for new ones, they were given away to the local population for home use as carpets, bed spreads and rags. The powdered smelter dross (solid wastes left after the lead extraction) was stored openly in the area belonging to the industry. It was given away without charge to the local population to pave their gardens and backyards, and also to the City Hall, to pave roads and recreation areas of local schools.

In 1980, the levels of Cd and Pb in the local vegetables and fruits, grown within a radius of 1,000 m, were determined as a function of the species and distance from the smelter [6]. The levels of both metals decreased with increasing distance from the smelter but no influence was observed as to the direction of the nominal prevailing winds. The dry basis concentrations of lead varied between 0.010 and 215 μg/g and those of cadmium between 0.004 and 11.8 μg/g for vegetables that had been pretreated and cleaned in a similar manner as for consumption. Greater concentrations were found in the leafly vegetables such as lettuce and mint, the edible part of which had a greater surface area in contact with the air. Lower concentrations were found in fruits with thick skins, such as banana and orange. However, considering the local dietary habits, precautions should be taken with regard to the constant consumption of okra, sweet potatoes, and cassava for both metals. The daily intake by a 60 kg human of more than 73 g of okra, grown at a distance of 200 m from the smelter, surpasses the provisional tolerable weekly intake of 7 μg Cd per kg body weight recommended by WHO/FAO [7]. In the same way, 150 g or more of sweet potato or cassava per day resulted in a higher weekly lead intake than the provisional tolerable level proposed by WHO/FAO (weekly limit for lead: 50 μg per kg body weight) [7].

In 1980, a biomonitoring program in the Santo Amaro area was set up by an interdisciplinary group[1] which later formed NIMA at the Federal University. The target population selected were 648 children aged 1–9 years living within 900 m from the main chimney of the smelter. Levels of Pb and Cd in blood were chosen as indicators of the absorption of these metals at the time of sampling and the level of zinc protoporphyrin (ZPP) in the blood was used as an indicator of the risk of intoxication. Levels of the two metals in hair were determined to verify its validity as an epidemiological index in such a situation. Clinical examinations were made by medical doctors who aimed to find symptoms of lead and cadmium intoxication, however, none were found [8]. The levels of the

---

[1] Carvalho FM, Silvany Neto A (medical doctors), Rocha, V (air chemist), Peso Aguiar M (zoologist), Tavares TM (analytical chemist and coordinator).

two metals as well as those of ZPP were very high compared with the corresponding levels from other parts of the world: lead blood concentrations (PbB) was in the range 0.77–7.30 μmol/L, with an average of 2.84 $\pm$ 1.2 [9]; this average is only below the level found in Silver Valley Study in Idaho, USA [10]; ZPP levels [11] varied between 0.07 and 14.45 μmol/L, with a geometric mean of 1.29 $\overset{\times}{\div}$ 2.36; about 4.7% of the children showed ZPP values higher than 4.81 μmol/L, which is the limit for the classification "extremely high" [12], and 38.4% had values above 2.11 μmol/L which is the limit for the classification considered "moderately high"; cadmium blood concentrations (CdB) varied between 0.004 and 0.51 μmol/L, with a geometric mean of 0.087 $\overset{\times}{\div}$ 2.5 [13]: these values were higher than any similar area with environmental contamination [14–16], or even unsimilar areas, where food was the main source of the contamination [17, 18]; lead concentrations in hair (PbH) varied between 20–4933 ppm, with a geometric mean of 313 $\div$ 3.2 [19, 20]; the cadmium level (CdH) varied between < 1–36 ppm, with a geometric mean of 4.1 $\div$ 2.2 [20, 21]; these values are higher than any corresponding area in the literature where the levels are due to environmental exposure [22, 23].

A technical report, based on the ZPP levels, was then written [24], by the University, and sent to the local authorities, who ordered the smelter to take the following recommended measures [25]: 1) Removal of the population living within 500 m of the smelter to other localities; 2) Take responsibility for the full treatment of the affected children; 3) Build a 90 m high stack to where all atmospheric emissions could converge[2]; 4) Install an efficient system of filtration for all sources of particulate matter from the smelter process; 5) Discontinue the donation of smelter dross and used filters; 6) Provide working clothing for all employees of the smelter (therefore, avoiding the transport of metals to the homes after working). All required measures were taken by the smelter except the relocation of all the people within 500 m which was reduced to a radius of 300 m. In this way, the smelter had to acquire an area of only 53,441 m^2 around the plant instead of an area of 493,265 m^2. The fullfillment of these requirements costed the industry US$ 163,365 plus a 50% reduction in production during two years.

In 1985 the same research group carried out a second cross-sectional study (again on their own initiative) on the children aged 1–9 years living at a distance 300–900 m from the chimney of the smelter. The average ZPP level decreased by 14.0%, PbB by 37.7%, and CdB by 67.8% [26,27]. In spite of this improvement, the levels of these metals in the blood remained high, with occurence of new cases of risk of lead intoxication [26, 27]. A new report was prepared for local authorities [28], at the cost of the University. Once more, the smelter plant was required to take the responsibility for the medical treatment of 31 children (12% of the total), considered as intoxicated and requiring medical care. No further requirements were imposed on the plant, although the government carried out

---

[2] This measure was conceived and specified by the environmental institutions of the state.

an inspection on the smelter and reported several irregularities with regard to secondary atmospheric emissions [29].

Further studies were then carried out by the same research group in an attempt to identify the causes of persistence of exposure to the two metals. The atmospheric concentrations of the two metals at a distance of 500 m from the smelter were still above the accepted levels for ambient air. At 900 m the cadmium levels exceeded some of the international standards such as the Swiss standard. The low dispersion potential, due to local geographical characteristics, plays an important role in this issue [27]. Fugitive emissions from the smelter, most probably from stocking, loading, unloading, and transportation of the ore is responsible for high levels of lead exposure to the population [27]. The presence of the smelter dross in the area around the homes accounts for a significant source of exposure to cadmium [27]. Additional mitigating measures need to be undertaken by the plant. A new report is being presently written and action will have to be taken in this direction.

In the following pages the role of different lead and cadmium reference samples, specially those of the blood in this case study will be discussed. Self-confidence for decision making was the most important advantage of its use.

Methodology

Atomic absorption with a graphite furnace has been used for the analysis of all biological matrices. ZPP in fresh blood was determined with a Buchler Hemafluor ZP. The analytical protocol used as well as the sampling designs and techniques applied are described elsewhere [3, 6, 8, 9, 11, 13, 21]. Details concerning analytical quality assurance are as follows:

In 1980, only one hemafluometer for ZPP measurements, within 2,000 km from Santo Amaro, was available. The methodology is very simple and specific. The only interference reported, of a biochemical nature, are iron deficiency or anemia, resulting in moderately elevated ZPP values, and a genetic disorder, erythropoietic protoporphyria, which is characterized by very high ZPP levels, but is very rare. To account for these possible interferences, the percentage of transferrin saturation (TS) as well as blood hemoglobin (HB) has been additionally determined for each blood, and statistical correction applied [8, 9, 11]. Clinical disturbances have also been checked. No reference samples are available for ZPP determination. As readings should be taken with fresh samples, no intercalibration was possible until a year later when the lead smelter purchased its own Buchler Hemafluor and carried out an intercalibration with the University, considered as satisfactory [30].

In the first survey, PbB was determined by the method described by Fernandez [31], in which the sample is diluted with 0.1% (w/v) Triton X-100[R] and 5 μl are injected directly into the graphite tube using pipet tips previously washed with xylene. As the deviations for repeated determination exceeded 10% in most cases blood dilution was carried out with 0.5% Triton X-100 (liquid scintillation grade) and satisfactory reproducibility was thus obtained. Because no reference samples for the lead level in blood were then available, the following steps for partial quality assurance were taken: 1) calibration curve was checked against EPA quality control samples for trace metal analysis and standard addition carried out for several samples of different concentration ranges; 2) orchard leaves certified sample SRM 1571 from the US National Bureau of Standards was analyzed by the same technician in charge of the blood analysis; 3) periodical determinations of the same blood, obtained from a healthy and not professionally exposed person (PbB = 7.0 μg/100 mL and ZPP = 20 μg/100 mL), were carried out during the entire period of analysis (ca. 5 weeks); 4) some blood samples, with different lead levels were intercalibrated with the Toxicology Laboratory of the London School of Hygiene and Tropical Medicine, London, using the same analytical methodology. From all these checking procedures, the one providing the best quality assurance of our data should have been the intercalibration exercise. However, all procedures were fully satisfactory (deviations < 10%) except for the intercalibration, where some of the results for the higher level blood samples (where consecutive dilutions were necessary to conform with the linearity range of the calibration curve) disagreed significantly; the results from the University being higher. This fact resulted in lack of self confidence for the research group and inhibited the use of these results.

During the second survey, PbB was determined by the method of Stoeppler et al. [32]. Control blood samples for metals were then available from Boehringer Mannheim GmbH. These samples, with concentrations of lead of 0.72, 2.08, and 3.76 μmol/L, were used during all analytical work and the results obtained were within the assigned confidence range.

Cadmium in blood was determined for both surveys by methodology described by Stoeppler and Brandt [33]. Because of insecurity caused by partially unsuccessful intercalibration for lead, blood samples taken in 1980 were kept in a freezer at $-18\,^{\circ}$C until 1982 when control blood samples for metals were made available. In addition to all other quality assurance procedures used for the PbB, accuracy of CdB was assured by checking against 48.9 and 168 nmol Cd/L blood control samples. The results obtained were always within the confidence range.

Hair samples have only been collected in 1980. The Pb and Cd determinations were carried out by ETAA, after washing each hair sample separately with 10% neutral Extran^R solution in ultrasound followed by digestion in HNO_3 [19, 20, 21]. No hair reference samples were available at that time. Hence, alternative procedures such as the use of NBS reference samples for other matrices (bovine liver and orchard leaves) as a control for general laboratory conditions, and daily determinations of an internal reference standard made from two hair samples, as a control of precision, had to be used. It was not until 1990 that a portion of the only existing hair reference sample, prepared in 1985, had been donated to us [34] and the values obtained were within the assigned confidence limits. Although anachronistical, this result increased the self-confidence of our research group.

Results and Discussion

From all the analytical data of the bioindicators used for monitoring of the children population in this study, the lead level in the blood from 1980 (PbB80) was the limiting factor for decision making. There had been no doubts on the results of ZPP, lead in blood from 1985 (PbB85), cadmium in blood from 1980 (CdB80), and cadmium in blood from 1985 (CdB85). No action was to be based on the hair data. The set of data from 1980, together with examples of similar reported studies in developed countries, plus common sense, constituted enough argumentation that measures recommended should be carried out by the smelter industry [24], except for the most urgent one at the time, that is, the nomination of the children to undergo treatment. The Center for Disease Control in the USA, CDC, was the institution which made more detailed recommendations on how to prevent lead poisoning in children [12]. They defined lead poisoning and proposed an interpretation of the screening results based on both tests, ZPP and PbB. Clinical management as to chelation therapy should involve a diagnostic evaluation. However, treatment has been re-commended, regardless of the presence or absence of clinical symptoms, for children with confirmed lead poisoning as defined by four criteria, one of them being PbB equal to or greater than $3.36\,\mu mol/L$ and another being $PbB > 2.35\,\mu mol/L$ with compatible symptoms and evidence of toxicity (for example, abnormal ZPP).

The values of ZPP and PbB80 of the children of Santo Amaro were among the highest reported in the literature, as already mentioned. However, the ratio ZPP/PbB for Santo Amaro children was very low when compared to the ones of other studies [11], particularly the one of Silver Valley [10]. This could be an indication of false high PbB80 results. Hence, only ZPP limiting criteria were used to define lead poisoning in Santo Amaro. This meant that children who ought to receive treatment, if PbB limiting criteria was also used, would not get it, because of the risk of giving unnecessary treatment to children not in need. In the 1985 cross-sectional study, where PbB values were obtained by blood control samples, the ratio ZPP/PbB was also low in relation to studies conducted elsewhere. This served to indicate that the results of PbB80 might not have been as erroneous as originally suspected. Table 1 presents the percentage of children exceeding the tolerance and critical limits of CDC, according to the criteria based on one test result as limiting factor. The values displayed show that 75% of the children exceeded the tolerance limit based exclusively on the PbB criteria, whereas only 38.4% exceeded this limit when defined by ZPP alone. This means that up to 36.6% of the children, which might have needed the chelating therapy, did not get it. If blood control samples had been available at that time, this would never have happened.

The smelter authorities and the state government never questioned the credibility of our data or the appropriateness of our recommendations. How-ever, doubts persisted within our research group, hindering and/or delaying the publication of the results involving data of PbB80 [9, 11, 27] or of hair [19, 20, 21]. At an international level the quality assurance of the analytical data of this

Table 1. Percentage of children exceeding the critical and tolerable limits of CDC, USA, for ZPP and PbB levels in the children population of Santo Amaro, during 1980 and 1985

CDC limits	% of population exceeding limit			
	1980		1985	
	ZPP	PbB	ZPP	PbB
Tolerance[a]	38.4	75	7.2	26
Critical[b]	4.7	25	0.8	4

[a] ZPP-2.11 to 4.79 μmol/L; PbB-2.35–3.35 μmol/L
[b] $\geqslant 4{,}80$ μmol/L; PbB $\geqslant 3.36$ μmol/L

study have sometimes been questioned, for example, at scientific meetings and during reviews of submitted papers to international journals, despite the described precautions of the analytical work. This is very often the case with data coming from developing countries, where laboratory conditions and expertise seldom meet the standards of developed countries. To avoid further uncertainties, international cooperation was sought in Europe, particularly with the Institut für Spektrochemie und Angewandte Spektroskopie, ISAS, in Germany. This resulted in the transfer of analytical know-how and inter-calibration of our samples, and the use of the now available blood and hair reference samples. Foreign financial support has enabled the purchase of some reference materials, additional training of local analytical chemists and the transportation of samples from and to Brazil.

Conclusions

Reference samples for different matrices are essential for data quality assurance and decision making, particularly in environmental issues. In developing countries, where fewer competent analytical laboratories exist, the relative importance of this is greater than in developed countries. In developing countries emission abatement systems have not been enforced in the past and technical systems for governmental emission control are still unsatisfactory; present cases involving high contamination tend to occur much more often than in developed nations. The cost/benefit ratio as to the use of reference material is thus lower for developing countries than for developed ones. The described case study of Santo Amaro is a good example of such a situation.

With the global approach to environmental issues, emphasized in the last few years, international investment for the development, production, and free distribution of environmental reference materials of geological and biological nature will bring results. Quality assurance of data from all parts of the world are essential for global data banks, balances, models and scenarios. Special international environmental programs for free reference samples of a geological and biological nature should be established in the near future; for example, the ASREM-Asian Society for Reference Material, which intends to offer reference samples, initially related only to food quality, to developing countries.

Acknowledgements: Recognition is here expressed to the following people for their support in obtaining the quality assurance data on Santo Amaro: Prof. Frank D'Itri, from Michigan State University, USA, for the EPA quality control samples for trace metal analysis; to the Rockefeler Foundation, USA, for the donation of NBS certified reference samples; to Dr. M. Stoeppler, from the Institute of Chemistry, Research Center, KFA-Jülich Germany, for the metal control blood samples; to Dr. K. Okamoto of the National Institute of Environmental Studies, Japan, for the hair reference samples and to the Toxicology Laboratory of the London School of Hygiene and Tropical Medicine, England, for the PbB80 intercalibration. Special thanks are addressed to the Institut für Spektrochemie und angew. Spektroskopie: to their directors, Prof. G. Tölg and Prof. D. Klockow, for their constant support, and particularly to Dr. F. Alt for collaboration in all steps of the quality assurance of the blood metal analyses.

References

1. Oliveira ER (1977) Parecer técnico sobre a ampliacão da COBRAC- Companhia Brasileira de Chumbo, em Santo Amaro, Bahia, Official Report CEPED/SEPLANTEC, Bahia (in Portuguese)
2. Bahia. SEPLANTEC.CPE (1976) Comportamento demográfico e divisão territorial do Estado da Bahia de 1940 a 1970. Volume 2. Salvador (in Portuguese)
3. Tavares TM, Carvalho FM, Peso-Aguiar MC (1983) Abstracts of the International Conference of Heavy Metals in the Environment
4. Carvalho FM, Tavares TM, Linhares P, Souza SP (1983) Cienc Cult 35 3: 360 (in Portuguese)
5. Carvalho FM, Tavares TM, Souza SP, Linhares PS (1984) Environ Res 33: 330
6. Petersen MNMB, Tavares TM (1981) Supl Cienc Cult 34: 539 (in Portuguese)
7. Joint FAO/WHO Expert Comm. Food Additives (1972) WHO Tech Rep Ser # 505 Geneva, Switzerland
8. Carvalho FM (1982) PhD Thesis, London School of Hygiene and Tropical Medicine, London
9. Carvalho FM, Barreto ML, Silvany-Neto AM, Waldrom HA, Tavares TM (1984) Sci Total Environ 31: 71
10. Landringan PJ, Baker EL, Feldman RG, Cox DH, Eden KV, Orenstein WA, Mather JA, Yankel AJ, Lindern IH (1976) J Pediatrics 89: 904
11. Carvalho FM, Silvany-Neto AM, Tavares TM, Lima MEC, Waldron HA (1985) PAHO Bulletin 19: 165
12. Center for Disease Control (1978) J Pediatrics 93: 709
13. Carvalho FM, Tavares TM, Silvany-Neto AM, Lima MEC, Alt F (1986) Environ Res 40: 437
14. Baker EL, Hayes CG, Landringan PJ, Handke JL, Leger RT, Housworth WJ, Harrington JM (1977) Am J Epidemiol 106: 261
15. Zielhius RL, Del Castilho P, Herber RFM, Wilbowo AAE, Salle HJA (1979) Int Arch Occup Environ Hlth 42: 231
16. Roels HA, Buchet JP, Lauwerys R, Bruaux P, Claesthreau F, Lafontaine A, Van Overschelde J, Verduyn G (1978) Environ Res 15: 290
17. Friberg L, Piscator M, Nordberg GF, Kjellstrom T (1974) Cadmium in the Environment, 2nd edn. CRC Press, Cleveland, Ohio

18. Thomas JFA (1979) Interim Report on Metal Contamination at Shipham, England
19. Tavares TM, Brandão AM, Chaves MEC, Silvany-Neto A, Carvalho FM (1989) Intern J Environ Anal Chem 36: 221
20. Carvalho FM, Silvany-Neto AM, Chaves MEC, Melo AMC, Galvão AL, Tavares TM (1989) Cienc Cult 41: 646
21. Carvalho FM, Silvany-Neto AM, Melo AMC, Chaves MEC, Brandão AM, Tavares TM (1989) Sci Total Environ 84: 119
22. Jenkins D (1979) Toxic Trace Metals in Mammalian Hair and Nails. Technical Report EPA-600/4-79-049
23. Hammer DI, Finklea JF, Hendricks RH, Shy CM, Horton RJM (1971) Amer J Epidem 93(2): 84
24. Carvalho FM, Tavares TM (1980) Estudo dos efeitos da poluição ambiental por chumbo e cádmio em uma população infantil. Technical Report. UFBa, Salvador, Bahia, Brazil
25. Governo do Estado da Bahia (1980) Decreto # 27.605. Salvador, Bahia, Brazil
26. Silvany-Neto A, Carvalho FM, Chaves MEC, Brandão AM, Tavares TM (1989) Sci Total Environ 78: 179
27. Tavares TM (1990) PhD Thesis, Universidade de São Paulo, Brazil
28. Carvalho FM, Tavares TM (1985) Technical Report. UFBa, Salvador, Bahia, Brazil
29. Aguiar J (1986) Relatório de Inspeção # 04/86. CRA-SEPLANTEC
30. Spínola A (1981) Personal communication
31. Fernandez FJ (1975) Clin Chem 21: 558
32. Stoeppler M, Brandt K, Rains TC (1978) Analyst 103: 714
33. Stoeppler M, Brandt K (1980) Frezenius Z Anal Chem 300: 372
34. Okamoto K, Morita M, Quan H, Fuwa K (1985) Clin Chem 31: 1592
35. G. Y. Iyengar (1990) Food Laboratory News 6(19): 20

5. Organic Analytical Approaches

5.1 Air Filter Systems – Necessary for Environmental Specimen Banking

J. Jacob and G. Grimmer

The following conclusions may be drawn from the findings presented in this paper:

- Passive samplers result in slightly different PAH profiles when compared to direct sampling systems. Nevertheless, there is a good chance to extrapolate from one matrix to another provided that these findings are confirmed by further investigations. This may qualify biological passive samplers for long-term monitoring of PAH and justify their integration in the German Environmental Specimen Bank (GESB).
- Among the passive samplers investigated, spruce sprouts were the optimum system.
- When compared to the main emission sources, the PAH-profile found in ambient air resembles most that of domestic coal heating exhaust which appears to be the main contributor to air pollution by PAH.
- Preliminary analyses carried out in the framework of GESB showed that there is a great difference in PAH-pollution between the western and eastern part of FRG.
- Within the past 6 years a significant decrease of air pollution by PAH was observed for the western part of FRG.

Introduction

Various epidemiological studies have shown that atmospheric pollutants may contribute to the health risk of humans. Hence, quality control of ambient air is essential. Correct air sampling and a reliable analysis of air pollutants, however, is associated with a series of problems which delayed the realization of such a project for quite a long time. Recently most of these problems have been satisfactorily solved so that the matrix 'air' is definitely represented now in the GESB. Problems and their solution will be discussed in this paper with regard to polycyclic aromatic hydrocarbons (PAH) as an example of one of the most important classes of environmental pollutants with a high carcinogenic potential.

Sampling Problems Using Active Collecting Systems

In principle four methods for the direct collection of ambient air samples are available: (1) direct sampling of air specimens; (2) freezing; (3) wash out by gas washing bottles, and (4) collection with suitable filter combinations (particle/volatile filter).

Specimen Banking
Rossbach/Schladot/Ostapczuk (Eds.)
© Springer-Verlag Berlin Heidelberg 1992

(1) Direct sampling of air specimens. Because of the low concentrations of PAH in ambient air (e.g. about 1–10 ng benzo(a)pyrene per m^3 air) direct sampling in gas sampling bulbs is unrealistic since sample sizes of several cubic meters would be required.

(2) Freezing. Condensation of the total air sample by freezing or separation of air constituents at low temperatures require complex and expensive equipment and, hence, cannot be recommended as a routine procedure.

(3) Wash out by gas washing bottles. Pollutants may be separated from air samples, in some cases even selectively (basic compounds by acidic treatment), by passing the air through gas washing bottles. The volatility of many compounds to be analyzed and of the solvent used for washing, however, often defeats a quantitative separation. In addition, large sample sizes have to be evaporated to small volumes prior to analysis when this sampling procedure is applied, which may result in significant losses of various low boiling constituents.

(4) Filter combinations. In principle, the most elegant method for air sampling is the collection with filter combinations although there are two severe problems to be overcome: (a) chemical destruction of sensitive compounds by reactive compounds in the air (b) the reevaporation of components already separated on the filter (blowing off effect). With regard to the PAH discussed here this means that especially those compounds possessing an exceptional high carcinogenic potential may readily react with e.g. SO_2, NO_x, oxygen or ozone and may escape the analytical determination. Reaction rates depend very much on the sort of adsorbents used; for instance PAH adsorbed on mineral materials are more rapidly converted than those adsorbed on carbonaceous material. This chemical destruction of PAH on the filter surface seems to play an important role in the case of PAH with anthracenoid partial structures such as benz(a)anthracene and dibenz(a,h)anthracene – compounds of high ecotoxicological relevance.

The volatility of low-boiling PAH such as phenanthrene, anthracene, fluorene, fluoranthene, and pyrene (boiling points $< 395\,°C$) results in high evaporation rates from commercially available glass fibre filters. These PAH are present in gaseous form in the atmosphere rather than adsorbed to particulate matter. For instance, Bjørseth et al. [1] found only 12 μg phenanthrene/m^3 in the particle but 450 μg phenanthrene/m^3 in the gaseous phase (ratio 1:38) of the atmosphere of an aluminium plant [1]. Broddin et al. [2] reported on annual variations of the distribution between the particle and the gaseous phase; they found higher concentrations in the gaseous phase (2:1 up to 80:1) during all periods.

Sampling with Filter Combinations

The high sensitivity of the detector system applied in the widely used glass capillary gas chromatography allows the reduction of the sample size to 1–3 m^3

and simultaneously minimizes the decomposition process of sensitive components on the filter by using collecting periods of only 10–60 minutes. A collecting device using a small low volume sampler (e.g. Kleinfiltergerät GS 050/3-N, Derenda, Berlin, FRG) and a filter combination consisting of a silicon-bound glass fibre and a Porapak PS filter have been previously described [3, 4]. A schematic presentation of the filter top is given in Fig. 1.

The separation of homologous compounds belonging to four chemical classes (n-alkanes, PAH, aromatic amines and azaarenes) with this filter system may be obtained from Tables 1–4 which present the distribution of the various compounds in both parts of the filter in per cent.

The tables also demonstrate that the retention on the two filter parts depends on the chemical nature of the compounds. n-Alkanes with boiling points > 300 °C are completely adsorbed by the glass fibre filter, whereas in the case of aromatic amines, azaarenes, and PAH only compounds with boiling points > 430 °C are completely absorbed by this filter.

Unfortunately, the above filter combination is not suitable for the collection of highly volatile constituents present in air samples (boiling points < 130 °C), because (1) not all of them are readily adsorbed completely on the filter material and (2) extraction with solvents is contraindicated because evaporation of the solvents results in more or less complete loss of the compounds which have to be analysed. The recommended analytical method in this case is direct evaporation of the sample from the filter material into the analysing system (GC/FID). This system is known as the thermodesorption procedure. Porapak PS (beads of a polymer of styrene/divinylbenzene), although a good adsorbent, cannot be used in this technique because of its poor thermostability. Among the various materials which have been checked so far, Tenax TA (poly(2,6-diphenyl-p-propylenoxide; 20–35 or 35–60 mesh) was found to be most promising. It may be seen from the VDI guideline 3482 [6] that breakthrough volumes for low-boiling air constituents and for polar compounds are about 1 l/g or less which makes sampling impossible in this case. Experiments carried out in our laboratory using Tenax TA (polyphenyl ether) confirmed this for several ecologically

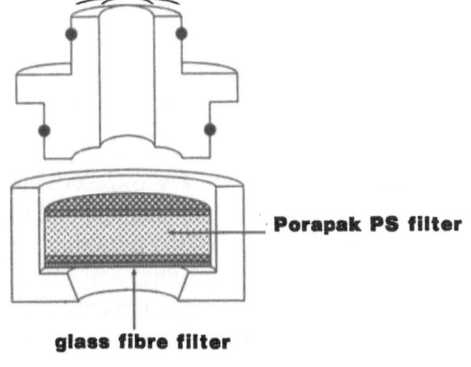

Porapak PS filter

glass fibre filter

Fig. 1. Filter combination for the collection of particulate matter and semivolatiles from ambient air (glass fibre filter: outer diameter 44 mm; Porapak filter: height 14 mm, inner diameter 44 mm)

Table 1. Distribution of *n*-alkanes from a spiked air sample to a Porapak PS particle filter and a glass fibre filter[a]

Compound	Boiling point (°C)	Percentage found in the Porapak filter	Glass fibre filter
n-Octane	126	100	0
n-Nonane	151	100	0
n-Decane	174	100	0
n-Undecane	196	98	2
n-Dodecane	216	96	4
n-Tridecane	234	97	3
n-Tetradecane	254	89	11
n-Pentadecane	270	62	38
n-Heptadecane	302	6	94
n-Octadecane	317	0	100
n-Nonadecane	330	0	100
n-Eicosane	220 (30 mm)	0	100

[a] *n*-Hexadecane has not been measured since it has been used as internal standard for GC; sampling conditions: 2.55 m³/h; 1 h sampling period; 44 mm filter diameter

Table 2. Distribution of various aromatic hydrocarbons from a spiked air sample to a Porapak PS particle filter and a glass fibre filter[a]

Compound	Boiling point (°C)	Percentage found in the Porapak filter	Glass fibre filter
Cumol	152	100	0
Mesitylene	165	100	0
Tetralene	208	100	0
Naphthalene	218	100	0
1-methylnaphthalene	245	100	0
Biphenyl	255	100	0
1,5-dimethylnaphthalene	269	100	0
Acenaphthene	278	100	0
Fluorene	298	100	0
Phenanthrene	338	61	39
Anthracene	340	68	32
Fluoranthene	384	13	87
Pyrene	394	17	83
Benzo(b)naphtho(2,1-d)thiophene	430	2	98
Benzo(e)pyrene	493	0	100
Benzo(a)pyrene	496	0	100
Indeno(1,2,3-cd)pyrene	534	0	100

[a] Conditions as in Table 1; indeno(1,2,3-cd)fluoranthene has been used as internal standard

relevant compounds. Combination filter systems – Tenax TA + Carbosieve S-III (active coal with a surface of 900 m²/g) – allow the collection of low boiling alkanes and alkenes even with sampling volumes of 10 l/g. A filter combination of Tenax TA + Carbosieve III + Carbotrap (highly pure graphite active coal with a surface of 100 m²/g) has been found to be advantageous for the collection of halogenated hydrocarbons, ketones, aldehydes, and alcohols. Poorer results

Table 3. Distribution of azaarenes from coke emission to a Porapak PS particle filter and a glass fibre filter (Grimmer et al. 1987 [5])[a]

Compound	Boiling point (°C)	Percentage found in the	
		Porapak filter	Glass fibre filter
4-azaphenanthrene	340	73	27
Acridine	344	37	63
Benz(c)acridine	434	3	97
Benz(a)acridine	438	3	97
dibenz(a,h)acridine	n.a.	0	100
dibenz(a,j)acridine	n.a.	0	100

[a] Conditions as in Table 1; 10-azabenzo(a)pyrene has been used as internal standard; n.a. = data not available

Table 4. Distribution of aromatic amines from coke emission to a Porapak PS particle filter and a glass fibre filter (Grimmer et al. 1987 [5])[a]

Compound	Boiling point (°C)	Percentage found in the	
		Porapak filter	Glass fiber filter
Aniline	184	79	21
o-toluidine	200	79	21
p-toluidine	201	78	22
m-toluidine	203	75	25
2-aminobiphenyl	299	24	76
1-aminonaphthalene	301	10	90
4-aminobiphenyl	302	31	69
2-aminonaphthalene	306	24	76
2-aminofluorene	n.a.	0	100
9-aminophenanthrene	n.a.	0	100
2-aminoanthracene	n.a.	0	100
3-aminofluoranthene	n.a.	0	100

[a] Conditions as in Table 1; 6-aminochrysene has been used as internal standard; n.a. = data not available

were obtained for *tert*-butylchloride, chloroform, acetone, and methylformiate under these conditions.

Passive Samplers

Pollutants may be transported from the matrix air into water, soil, and plants (leaves) by sedimentation of air particulates, atmospheric washout, and by gas–liquid distribution. Hence, another approach to air sampling is the application of passive sampling systems. It seems that leaves, spruce sprouts, and grass are suitable matrices for passive sampling. An accumulation of lipophilic compounds into the cuticular wax layer of plants has been observed and recommend them as passive samplers. In autumn, at the end of the vegetation period, this material becomes part of the soil and leads to a considerable accumulation of PAH in this matrix.

In the framework of the GESB, PAH contents of leaves from beech and poplar trees, of spruce sprouts, and of soil originating from various areas of FRG have been determined and compared to those present in air samples collected from these areas. The results of this study are described in the following six paragraphs.

Range of Local Variations of the PAH-Concentration in a Sampling area using Spruce Sprouts as Passive Sampler

Variation coefficients of $+/-50\%$ were found for the PAH-concentrations in spruce sprouts collected in ten areas of the Bavarian Forest from a total of three trees (n = 30). The fairly large range is reduced to about $+/-20\%$ when PAH-profiles are considered in which the concentrations of the various PAH have been related to benzo(e)pyrene (BeP = 1.00). The comparatively large range of the absolute concentrations may be explained by the fact that they were related to fresh weight and that the material was still heterogenous in some cases (different wood content). The data are condensed in Fig. 2.

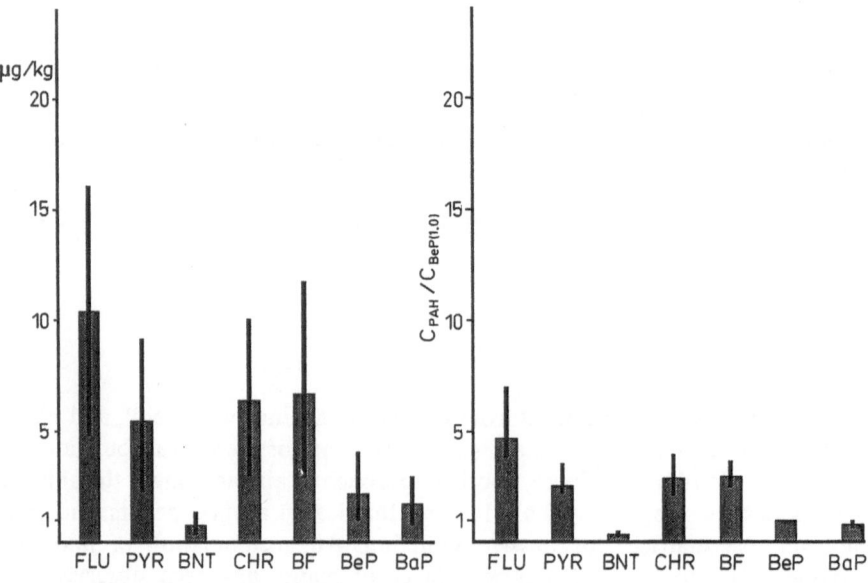

Fig. 2. Range of the PAH-concentration measured in spruce sprouts from a distinct area (Bavarian Forest, 1990)
FLU (fluoranthene); PYR (pyrene); BNT (benzo[b]naphtho[2,1-d]thiophene); CHR (chrysene); BF (benzofluoranthenes); BeP (benzo[e]pyrene); BaP (benzo[a]pyrene)

PAH-Profiles in Various Soil Horizons

Prior to attempts to correlate PAH-profiles from soil and plant material to air samples, the dependence of these profiles on the depth from which soil samples were taken was studied. Four areas were selected (Hirschbichltal 1 and 2, Saarbrücken-Warndt (all in 1989)), and four horizons were investigated: O (organic horizon characterized by its humus accumulation); A (mineral surface horizon with accumulation of organic material); B (mineral subsurface horizon); M (mineral horizon). The PAH-concentration of a specific layer in a distinct area (e.g. Hirschbichltal) was found to be almost constant. As expected, great differences were found between soils collected in areas differently exposed to air pollution. Table 5 presents some concentrations of benzo(e)pyrene found in various soil horizons taken from clean air (Hirschbichltal) and an industrialized area (Saarbrücken-Warndt).

Although the concentrations found in these two areas differ by one order of magnitude the PAH-profiles obtained were very similar. Profiles were found, however, to depend on the depth from which samples were taken. The ratio of the concentration of lower boiling to higher boiling PAH decreases significantly from the surface (O) to the mineral horizon (B or M). This may indicate a biological decay of PAH with lower molecular weights (fluoranthene, pyrene, chrysene (Fig. 3)).

Correlation of PAH-Concentrations found in Air, Plant, and Soil Samples Collected in the same Area

For some areas PAH-concentrations have been measured in matrices which depend on each other (air, poplar and beech leaves, spruce sprouts and soil

Table 5. Benzo(e)pyrene concentration in soil samples of various horizons taken from two different areas

Origin	BeP concentration (μg/kg)
Hirschbichltal (Berchtesgaden)[a]	
Area I	
O-horizon	22.4
A-horizon	27.0
M-horizon	5.4
Area II	
O-horizon	18.4
A-horizon	16.3
M-horizon	2.3
Saarbrücken-Warndt[b]	
O-horizon	468
A-horizon	758
B-horizon	253

[a] Clean air area;
[b] Industrialized area

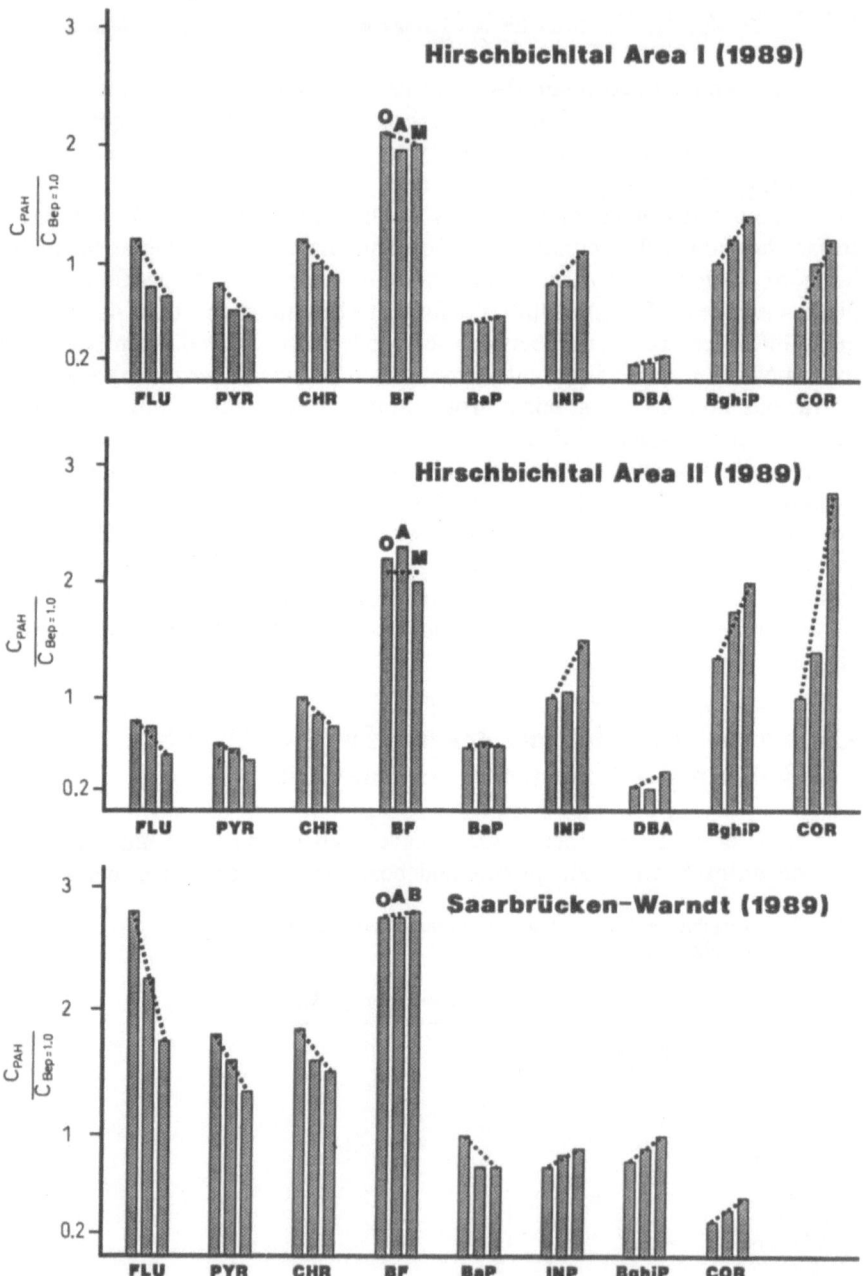

Fig. 3. PAH-profiles in dependence of the soil horizon (Hirschbichltal Area I and II, Berchtesgaden and Saarbrücken-Warndt, April 1989)

PAH-profiles in dependence of the soil horizon (Hirschbichltal Area II, Berchtesgaden; April 1989)

PAH-profiles in dependence of the soil horizon (Saabrücken-Warndt, 1989)

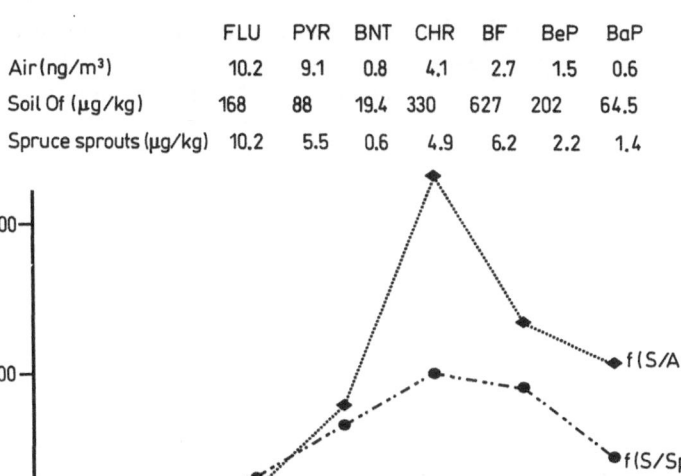

	FLU	PYR	BNT	CHR	BF	BeP	BaP
Air (ng/m³)	10.2	9.1	0.8	4.1	2.7	1.5	0.6
Soil Of (µg/kg)	168	88	19.4	330	627	202	64.5
Spruce sprouts (µg/kg)	10.2	5.5	0.6	4.9	6.2	2.2	1.4

Fig. 4. PAH-concentration in air, spruce sprouts and soil from the Bavarian Forest (23.4.1990) and correlation factors

$f_{S/A}$ = factor to extrapolate PAH-concentration from soild to air

$f_{S/Sp}$ = factor to extrapolate PAH-concentration from soild to spruce sprouts (for abbreviations of PAH see Fig. 2)

(O-horizon). An example is presented in Fig. 4 (Bavarian Forest) which shows that there is at least no simple correlation. Correlation factors e.g. for extrapolation from soil to air ($f_{S/A}$) or from soil to spruce sprouts ($f_{S/Sp}$) differ from one PAH to another.

Correlation of the PAH-Profiles found in Air, Plant, and Soil Samples Collected in the same Area

Better correlation, however, is found when PAH-profiles of these matrices are compared (see Fig. 5). Mathematical analysis of the numeric PAH-concentrations (C) and their profiles results in a correlation which may be expressed as:

$$C_{air(ng/m^3)} = \frac{C_{plant\ leaf\ (ug/kg)} \times f_{(PAH)}}{1.47}$$

Fig. 3. Continued

FLU (fluoranthene); PYR (pyrene); CHR (chrysene); BF (benzofluoranthenes); BaP (benzo[a]pyrene); INP (indeno[1,2,3-cd]pyrene); BghiP (benzo[ghi]perylene); COR (coronene)

O = organic horizon characterized by its humus accumulation

A = mineral surface horizon with accumulation of organic material

B = mineral subsurface horizon

M = mineral horizon

using factors for the various PAH which also depend on the kind of plant material (beech leaves, poplar leaves, spruce sprouts) as given in Table 6. This correlation approach, however, is very preliminary and requires confirmation by further experiments.

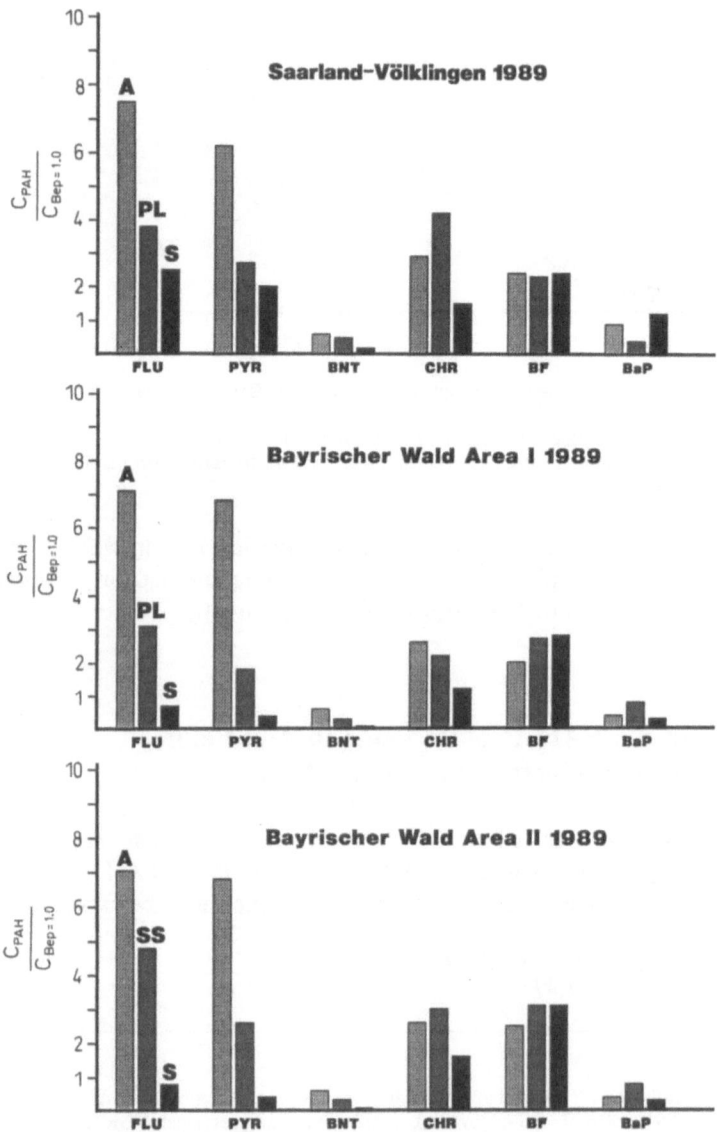

Fig. 5. Comparison of the PAH-profiles (C_{PAH}/C_{BeP}) found in air, polar leaves, spruce sprouts and soil samples
A = air; PL = poplar leaves; S = soil; SS = spruce sprouts

Table 6. Correlation factors to extrapolate PAH-concentrations in air and plant material

PAH	f Beech leaves	f Poplar leaves	f Spruce sprouts
Fluoranthene	2.22	1.97	1.45
Pyrene	3.53	2.30	2.58
Chrysene	1.14	0.69	0.87
Benzofluoranthenes (b + j + k)	0.74	1.00	0.65
Benzo(a)pyrene	0.63	0.90	0.56
Benzo(e)pyrene	1.00	1.00	1.00

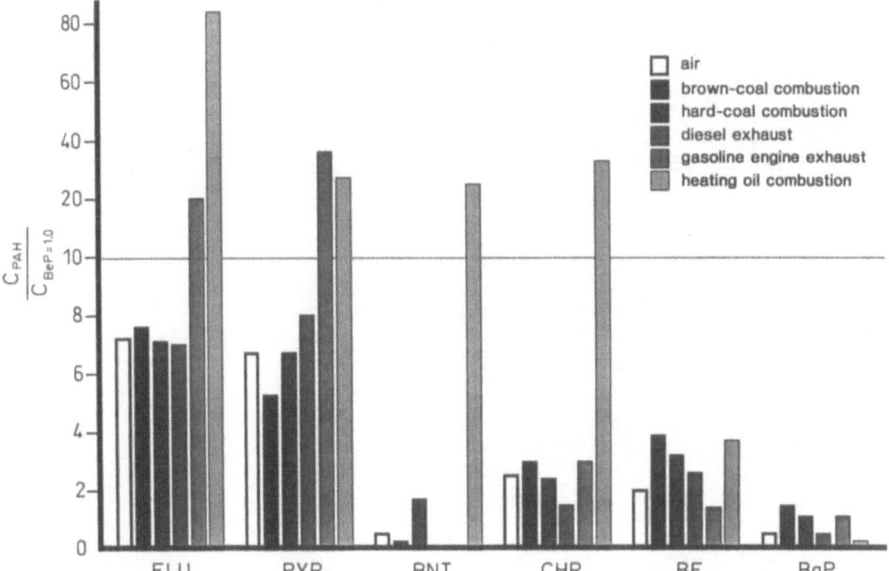

Fig. 6. PAH-profiles of the most important sources of emission in the FRG in comparison to ambient air

Surprisingly, fairly constant PAH-profiles were found in air samples from different areas of the country (Bavarian Forest, Saarland-Völklingen). This may indicate that one predominant emission source is mainly responsible for air pollution by PAH. Among the various emissions in question, hardcoal combustion exhaust exhibits the most similar PAH-pattern so that it may be assumed that this source plays the most important role for the PAH burden of ambient air in FRG (see Fig. 6).

Differences between the Various Biological Passive Samplers Tested

All three matrices tested as biological passive samplers (beech and poplar leaves, spruce sprouts) resulted in *qualitatively* similar PAH-profiles. At least in the range of the higher boiling compounds there are also *quantitative* correlations. Spruce sprouts have been found to be more suitable for collecting PAH, especially for the lower boiling ones for which less reliable results were obtained when the other systems were applied. But even with spruce sprouts the actual

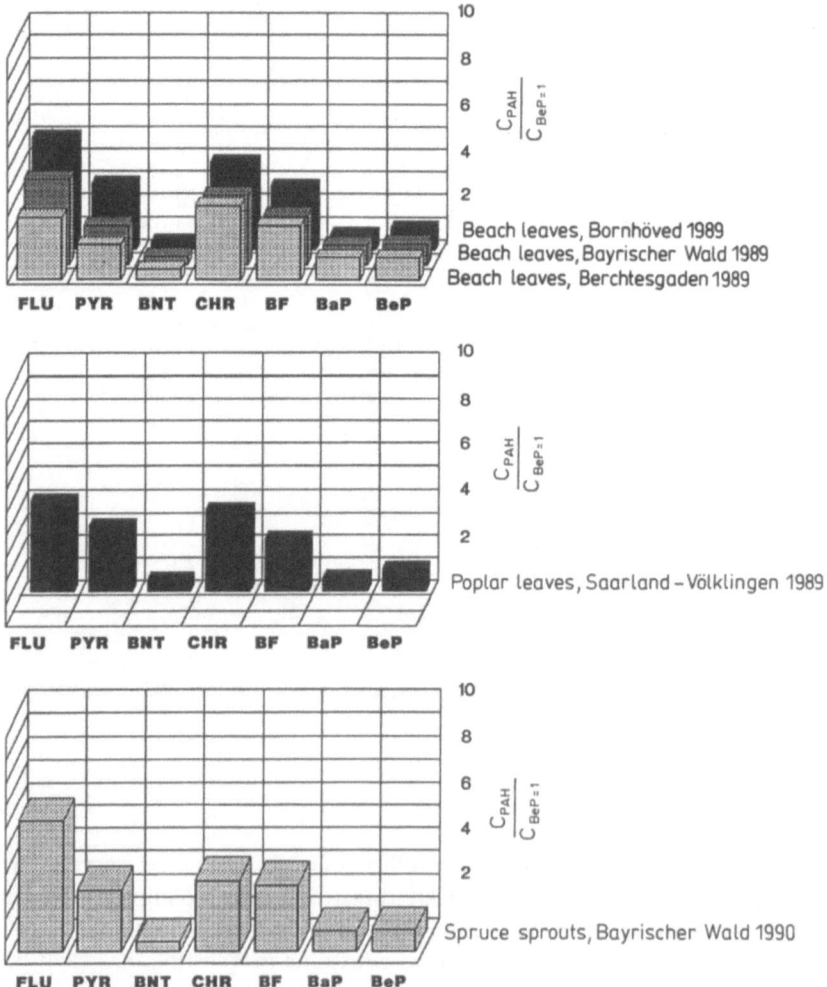

Fig. 7. PAH-profiles found in beech and poplar leaves and in spruce sprouts (for abbreviations of PAH see Fig. 2)

situation of ambient air is not correctly reflected. A comparison of three matrices is presented in Fig. 7.

Benzo(a)- and Benzo(e)pyrene Concentration as a Marker for Air Quality Control in the FRG over a Period of six Years

In two areas of the FRG (Saarland-Warndt; Berchtesgaden-Hirschbichltal) PAH have been measured in spruce sprouts as passive samplers over a period of six years. As shown for benzo(a)- and benzo(e)-pyrene in Fig. 8 the concentration decreased to about 50% of the initial values indicating a remarkable improvement of the air quality in FRG.

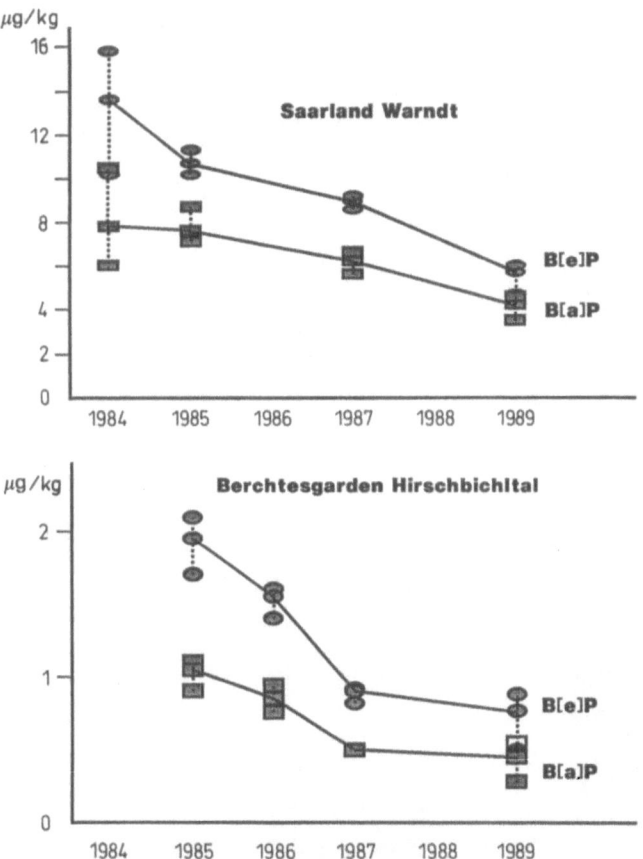

Fig. 8. Benzo(a)- and benzo(e)pyrene concentration in spruce sprouts from two areas of FRG over a period of 6 years

Table 7. PAH-concentration in ambient air from eastern and western parts of FRG in 1990 as measured by active sampling with combination filters (ng/m^3)

	MV	BF	SVW	SVH
Fluoranthene	114	9.6	13.6	36
Pyrene	68.7	8.6	11.2	28
Benzo[b]naphtho[2,1-d]thiophene	4.8	0.7	0.7	1.8
Benz[a]anthracene	12.3	1.3	2.7	7.5
Chrysene + triphenylene	15.7	3.1	5.1	12.6
Benzofluoranthene [b + j + k]	11.0	2.5	4.2	16.8
Benzo[e]pyrene	3.4	1.2	1.8	8.2
Benzo[a]pyrene	3.0	0.5	1.5	7.0

MV = Mecklenburg-Vorpommern
BF = Bavarian Forest
SVW = Saarland; Völklingen-Warndt
SVH = Saarland; Völklingen-Hostenbach

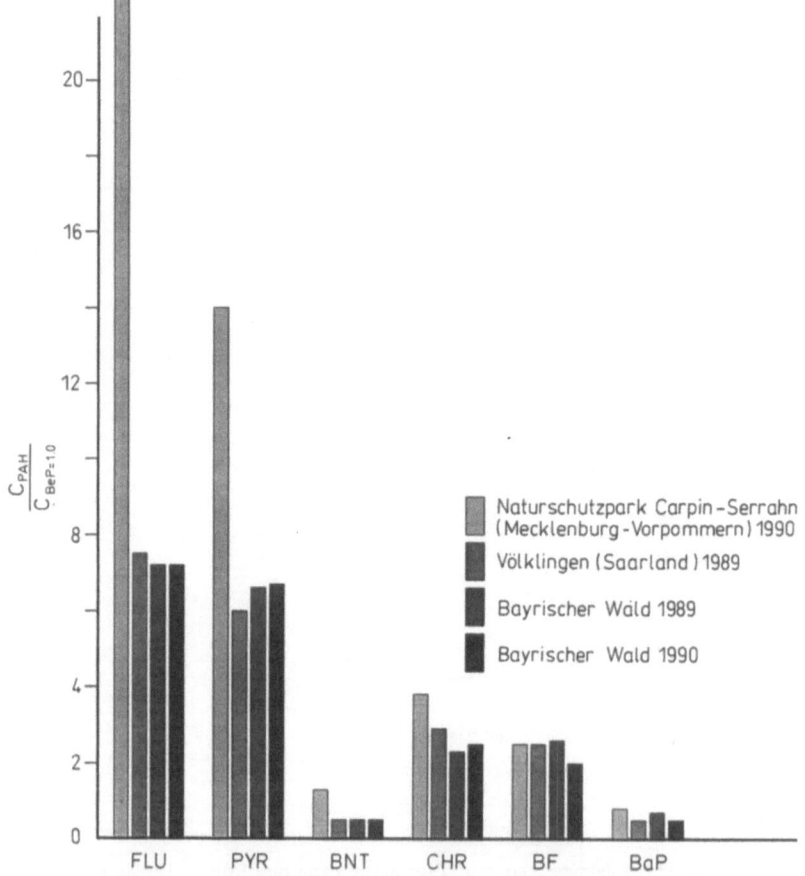

Fig. 9. PAH-profile in air samples (glass fibre- + Porapak PS-filter) (related to B[e]P = 1.00)

Recent measurements in rural areas of the eastern part of FRG using active sampling systems resulted in about ten times higher PAH concentrations than were found in comparable rural western parts of the country. Some of them were still about three times higher than those measured in industrialized areas of western FRG (Table 7).

Apart from that, slightly different PAH-profiles were found in the two areas with considerably higher concentrations of the lower boiling PAH found in samples from eastern FRG (Fig. 9).

References

1. Bjørseth A, Bjørseth O, Fjeldsted PE (1978) J Work Environ Health 4: 224
2. Broddin G, Cautreels W, van Cauwenberghe K (1980) Atmos. Environ 14: 845
3. Grimmer G, Naujack KW, Dettbarn G (1987) Toxicol. Lett 35: 117
4. Jacob J, Grimmer G, Schneider D (1990) Fresenius Z Anal Chem. 337: 73
5. Grimmer G, Naujack KW, Dettbarn G (1987) Beitrag zur Ursachenforschung exogen bedingter Blasencarcinome – Profilanalyse aromatischer Amine am Arbeitsplatz. Schriftenreihe der Bundesanstalt für Arbeitsschutz FG 511. DFVLR (Ed.) 39p.
6. VDI-Richtlinie 3482 Bl.6 Messen gasförmiger Immissionen. Gaschromatographische Bestimmung organischer Verbindungen – Probenahme durch Anreicherung – Thermische Desorption 1986

5.2 A Simple Cleanup Procedure for the Quantitative Determination of PAHs and Nitro-PAHs in Particulate Matter

K. Reisinger, A. Zajc

A single-step cleanup technique for the quantitative determination of polycyclic aromatic hydrocarbons (PAHs) and Nitro-PAHs is presented. The basis for the cleanup procedure is a solid–liquid chromatography on a column containing two silica gel packings of different mesh sizes. The eluents provide suitable fractions for identification and quantitation of PAHs and Nitro-PAHs by high resolution gas chromatography (HRGC) with a flame ionization detector (FID) and an electron capture detector (ECD), respectively. Analytical data obtained from investigations of NBS Standard Reference Materials 1649 and 1650, and a fly ash, reveal that the procedure is applicable without further modification to different complex samples.

Introduction

The global distribution of polycyclic aromatic hydrocarbons (PAHs) and nitro-polycyclic aromatic hydrocarbons (Nitro-PAHs) has been well established by extended investigations of environmental samples by means of organic trace analysis [1–5].

The quantitative determination of toxic PAHs and Nitro-PAHs (some of them are classified to be mutagenic and carcinogenic) is commonly performed by high resolution gas chromatography (HRGC): PAHs with a flame ionization detector (FID) [6] or by high performance liquid chromatography (HPLC) with UV- and fluorescence detection [7], and Nitro-PAHs by HRGC (ECD, FID) and HPLC [8]. Gas chromatography-mass spectrometry (GC-MS) for quantitation is applied to all three compound classes [8, 9]. For the separation of PAHs and Nitro-PAHs from accompanying substances a great variety of cleanup procedures is described in the literature [10–19, 20–31].

The work presented here describes a simple and rapid single-step cleanup procedure for the separation of PAHs and Nitro-PAHs from interfering substances. It is based on the method published recently for the separation of PCBs [32]. The applicability to extracts of different matrices is demonstrated by the quantitative and simultaneous determination of PAHs and Nitro-PAHs in NBS Standard Reference Materials 1649 and 1650, and a fly ash collected during the combustion of fuel oil.

Additionally, a contribution reporting a method for the quantitative determination of PAHs in air and air particulate matter using the cleanup procedure

Specimen Banking
Rossbach/Schladot/Ostapczuk (Eds.)
© Springer-Verlag Berlin Heidelberg 1992

introduced here, was presented at the 13th PAH Symposium held in Bordeaux, France, October 1990 [33].

Experimental

Chemicals and Solvents: *N*-hexane and dichloromethane of analytical grade were distilled in a fractionation column before use. All chemicals were purchased from Merck, Darmstadt (Germany). Toluene was a "brand high purity solvent" manufactured by American Burdick & Jackson and distributed by Fluka, Neu-Ulm (Germany).

Column Materials: Untreated silica gel 60 (particle size 70-230 mesh, water content 5.6%) and silica gel 70 (particle size < 230 mesh, water content 6.3%) served as the solid phases for the cleanup procedure. Both are available from Merck, Darmstadt (Germany).

Glass-ware: Micro-Soxhlet-extractors with glass filter frits for solid-liquid extraction of samples materials were purchased from Normag, Hofheim am Taunus (Germany). One-way Pasteur pipettes and one-way micropipettes used for the cleanup procedure and the preparation of solutions, respectively, were from Brand, Wertheim (Germany). *N*-hexane fractions from the cleanup procedure were collected in minivials with tapered, conical interior, Teflon-faced seals and graduated from 0.1 to 5 ml. These vials were purchased from Chrompack, Frankfurt-Niedereschbach (Germany). All glass equipment was cleaned with chromic–sulfuric acid, washed with bidistilled water, rinsed with acetone and *n*-hexane and stored at 250 °C until it was used.

Standard Solutions

PAHs: Stock solutions of PAHs in cyclohexane are made from solid compounds purchased from Promochem (Wesel, Germany), Aldrich (Steinheim, Germany) and Merck (Darmstadt, Germany).

The concentrations of the stock solution ranged from 45 mg pyrene/50 ml to 1.07 mg coronene/50 ml. The three final calibration standards contained 20 PAHs: fluorene, anthracene, phenanthrene, fluoranthene, pyrene, 1-methylpyrene (internal standard, i.st.), benz(a)anthracene, chrysene, 2,2'-binaphthyl (i.st.), benzo(b)fluoranthene, benzo(k)fluoranthene, benzo(e)pyrene, benzo(a)pyrene, perylene, indeno(1,2,3-cd)pyrene, dibenzo(ah)anthracene, benzo(b)chrysene (i.st.), benzo(ghi)perylene, anthanthrene and coronene with concentrations ranging from 3.0 ng, 1.2 ng, 0.6 ng coronene/μl to 18.1 ng, 7.25 ng and 3.62 ng pyrene/μl.

Nitro-PAHs: Nitro-PAH stock solution in cyclohexane is prepared from solid materials available from the above mentioned trading companies. The stock solution was a mixture of nine Nitro-PAHs: 1-nitronaphthalene, 2-nitrofluorene, 1-nitropyrene, 6-nitrochrysene, 3-nitrofluoranthene, 1-nitroperylene, 6-nitrobenz(a)pyrene and 9-nitroanthracene. The concentrations of the standards prepared from stock solution ranged from 26.5 pg, 13.2 pg, 6.6 pg, and 3.3 pg 3-nitrofluoranthene/μl cyclohexane as the lowest to 95.2 pg, 47.6 pg, 23.8 pg and 11.9 pg 6-nitro-benz(a)-pyrene/μl cyclohexane as the highest.

Sample materials: The preparation and cleanup procedure developed for the rapid separation of PAHs and Nitro-PAHs in complex samples was tested with 2 standard reference materials and a real fly ash sample. The latter was collected on a glass fibre filter during the combustion of fuel oil in the presence of a stationary atmosphere with a circulating fluidized bed combustor [34, 35]. The standard reference materials were:

1. NBS Reference Material 1649 (SRM 1649) – Urban Dust.
2. NBS Reference Material 1650 (SRM 1659) – Diesel Particulate Matter

The procedure has also successfully been applied as the cleanup step in the course of the quantitative determination of PAHs in biological and environmental materials such as copepod, mussel and sea plant homogenate and marine sediment [36].

High Resolution Gas Chromatography (FID): Gas chromatographic analysis for the quantitation of PAHs was performed on a Carlo Erba gas chromatograph, Model 2900, equipped with a flame ionization detector (FID) and a split-splitless injector. The GC column was a 30 m × 0.32 mm i.d. fused silica capillary column, film thickness was 0.25 μm (DB=5, J & W Scientific, Folsam, CA 95630, USA). The *operating conditions* were as follows: injector 320 °C, detector 340 °C; column initial temperature 95 °C, hold for 2 minutes; increase temperature 10 °C/min to 220 °C, hold 1 minute and increase 40 °C/min to 320 °C. Carrier gas was He at 70 kPa, H_2 at 50 kPa, and synthetic air at 110 kPa. 1 μl was injected splitless (valve opened after 1 minute).

High Resolution Gas Chromatography (ECD): Gas chromatographic analysis for the quantitation of Nitro-PAHs was done on a Carlo Erba gas chromatograph, Model 4160, equipped with a ^{63}Ni-electron-capture-detector (ECD) and a cold-on column injector. The GC-column was 50 m × 0.32 mm i.d. fused silica capillary column, film thickness was 0.25 μm (FS-SE 54 CB, Macherey & Nagel, Düren, Ger). The *operating conditions* were as follows: detector temperature 350 °C; temperature program for Nitro-PAH detection: initial temperature 100 °C; heat ballistically to 180 °C and with 5 °C/min to 320 °C, hold the upper temperature for 10 min. Carrier gas was He at 150 kPa; make-up gas was N_2 at 80 kPa. 1 μl was injected cold-on-column.

Sample Preparation

Soxhlet Extraction Procedure: 0.5 g samples or 500 mg samples of SRM 1649, 30 mg of SRM 1650 (the NBS Standards were taken without pretreatment) and the filters loaded with the fly ash weighed into glass filter frits were spiked with 1 μg of the internal standards for PAH quantitation and extracted 12 h with 20 ml of different solvents: SRM 1649 four times with *n*-hexane, cyclohexane, toluene and a mixture of cyclohexane/dichloromethane (4:1 = v:v), the fly ash samples four times and SRM 1650 three times with the cyclohexane/dichloromethane for PAH and Nitro-PAH determination. Photodecomposition of PAHs and Nitro-PAHs was avoided by using extractors covered with aluminium foil. The 20 ml portions of solvents were carefully concentrated to 300 μl with a stream of nitrogen at room temperature and the concentrate submitted to the cleanup procedure for removal of interferences.

Cleanup Procedure

Column Preparation: The column preparation procedure was reported elsewhere [32].

Cleanup: An appropriate volume (up to 300 μl) of the concentrated extract was directly applied to the silica-gel column (which was covered with aluminum foil) with a micropipette. Prior to cleanup the toluene extract was very gently evaporated to dryness at room temperature, the residue dissolved in *n*-hexane or cyclohexane and administered onto the column. After the first fine-meshed silica gel had taken up the concentrate 5.5 ml *n*-hexane were added in small portions. This fraction can be discarded unless PCBs are to be determined [32]. PAHs were eluted with 3.5 ml of a slightly polar mixture of *n*-hexane/dichloromethane (4.1 = v:v, fraction 2) and Nitro-PAHs with a more polar mixture of 5.5 ml *n*-hexane/dichloromethane (2:1 = v:v, fraction 3).

The purified fractions containing the compounds of interest were taken without further treatment or after concentration for analysis by gas chromatography (or gas chromatography-mass spectrometry).

Results and Discussion

Calculation Method for PAH and Nitro-PAH quantitation

Gas chromatography: GC (FID) and GC (ECD) were used for the quantitative determination of PAHs and Nitro-PAHs, respectively. For PAHs quantitation was achieved according to the internal standard method. Nitro-PAH concentrations were calculated by means of calibration curves established for 8 Nitro-PAHs of four different concentrations (see chapter "Standard solutions").

NBS Standard Reference Material 1649 – Urban Dust

PAH Concentrations: The comparison of the PAH concentrations in the reference materials with values given by NIST (Table 1) reveals that, due to the

Table 1. PAH concentration in Standard Reference Materials 1649 (urban dust) in comparison with certified values specified by NBS

Extraction Solvent	Fluoranthene [µg/g]	[%]	Benz(a)anthracene [µg/g]	[%]	Benzo(a)pyrene [µg/g]	[%]	Benzo(ghi)perylene [µg/g]	[%]	Indeno(1,2,3-cd)pyrene [µg/g]	[%]
Certified Values	7.1 ± 0.5^b		2.6 ± 0.3		2.9 ± 0.5		4.5 ± 1.1		3.3 ± 0.6	
n-Hexane	3.0 ± 0.78 (n = 4)	$42^a \pm 11$	1.4 ± 0.26 (n = 4)	54 ± 10	1.2 ± 0.50 (n = 4)	41 ± 17	2.1 ± 0.64 (n = 4)	47 ± 11	2.4 ± 0.48 (n = 4)	73 ± 15
Cyclo-hexane	4.4 ± 0.36 (n = 4)	63 ± 5	2.0 ± 0.59 (n = 4)	77 ± 23	2.0 ± 0.20 (n = 4)	69 ± 7	2.6 ± 0.29 (n = 4)	58 ± 7	2.6 ± 0.26 (n = 4)	79 ± 8
Toluene	5.3 ± 0.35 (n = 4)	74 ± 5	1.9 ± 0.05 (n = 4)	72 ± 2	2.2 ± 0.13 (n = 4)	76 ± 4	3.5 ± 0.21 (n = 4)	78 ± 5	3.3 ± 0.25 (n = 4)	100 ± 8
Cyclohexane Dichloro-methane (4:1)	4.8 ± 0.08 (n = 5)	68 ± 1	1.9 ± 0.07 (n = 5)	73 ± 3	2.1 ± 0.21 (n = 5)	72 ± 7	3.3 ± 0.43 (n = 5)	74 ± 10	3.1 ± 0.28 (n = 5)	94 ± 9

[a] Recovery rate (NBS values defined as 100%);
[b] Standard deviation

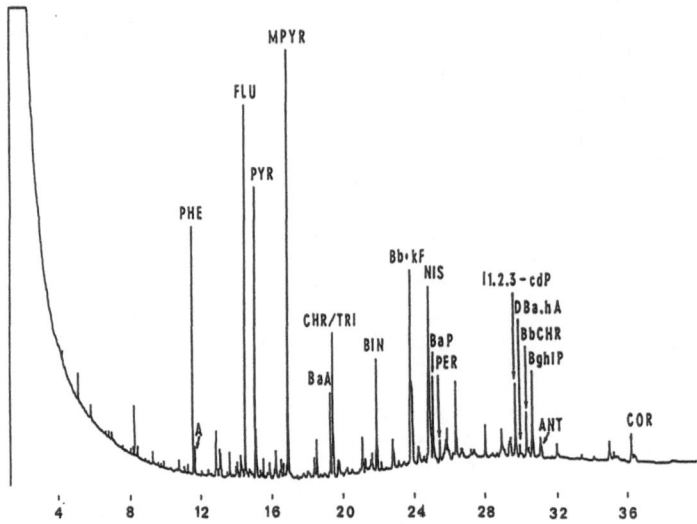

Fig. 1. GC (FID) chromatogram of a PAH mixture isolated from NBS Standard Reference Material 1649 (Urban Dust)

high solubility for PAHs, toluene shows the best mean recovery rates and in fact toluene seems to be the solvent of choice. However, it can be seen from Table 1 that extraction with a mixture cyclohexane/dichlormethane gave similar recoveries to those obtained with toluene. Therefore, the authors decided to use

Table 2. PAH concentrations found in Standard Reference Material 1649 (urban dust) – uncertified values specified by NBS

PAHs	NBS Values[a] [μg/g]	Mean Recovery Rate [%]
PHE[b]	4.5 ± 0.3[e]	58 ± 4
A	–	0.3 ± 0.07[c]
PYR	7.2 ± 0.2	52 ± 2
CHR/TRI	3.5 ± 0.1	89 ± 3
BbF	6.2 ± 0.3	95 ± 4[d]
BKF	2.0 ± 0.1	
BeP	3.3 ± 0.2	97 ± 9
COR		2.2 ± 0.5[c]

[a] NBS values defined as 100%;
[b] Abbreviations used: A = Anthracene, PHE = Phenanthrene, FLU = Fluoranthene, PYR = Pyrene, BaA = Benz(a)anthracene, CHR = Chrysene, TRJ = Triphenylene (not separated from CHR), BbF = Benz(b)fluoranthene, BkF = Benz(k)fluoranthene, BeP = Benz(e)pyrene, BaP = Benz(a)pyrene, PER = Perylene, I1,2,3,-cdP = Ideno-(1,2,3-cd)pyrene, DiBahA = Dibenz(ah)anthracene, BghiP = Benz(ghi)perylene, COR = coronene;
[c] Value in μg/g, not stated by NBS;
[d] Values represent the sum of BbF and BkF;
[e] Standard deviation (n = 5)

the mixture of cyclohexane/dichloromethane (4:1 = v:v) for all PAH and Nitro-PAH determinations.

The concentrations of PAHs extracted from NBS Standard Urban Dust by means of cyclohexane/dichloromethane are listed in Table 2 and compared with those given by the NBS. Figure 1 shows the gas chromatogram of the PAH mixture isolated from Urban Dust (for abbreviations see annotations of Table 2).

NBS Standard Reference Materials 1650 – Diesel Particulate Matter

PAH and Nitro-PAH Concentrations: The NBS Standard Reference Material 1650 was extracted for PAHs and Nitro-PAHs with cyclohexane/dichloromethane (4:1 = v:v). The PAHs were quantitatively determined by GC (FID) and the Nitro-PAHs by GC (ECD). The concentrations found in diesel particulate matter are summarized in Table 3 taking NBS values as 100%. Figures 2 and 3 show the corresponding gas chromatograms of the PAH and Nitro-PAH compounds.

Fly Ash Material Collected During the Combustion of Fuel Oil

PAH and Nitro-PAH concentrations: Teflon-coated glass fibre filters loaded with fly ash from the combustion of fuel oil were extracted and prepared using

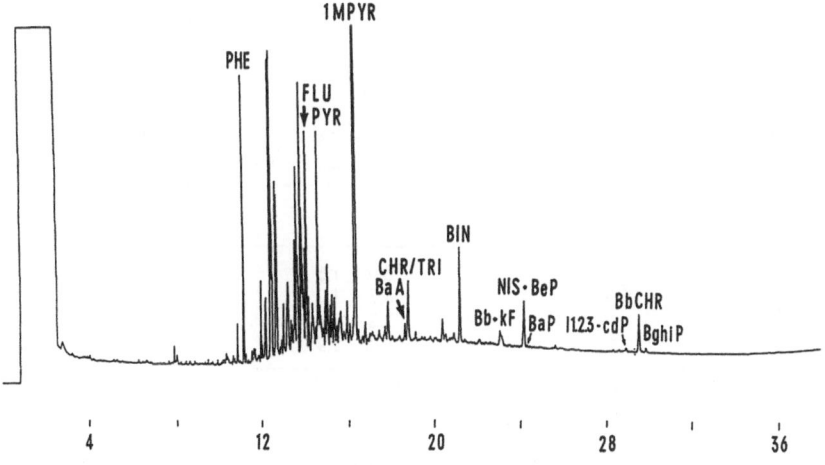

Fig. 2. GC (FID) chromatogram of a PAH mixture isolated from NBS Standard Reference Material 1650 (Diesel particulate matter)

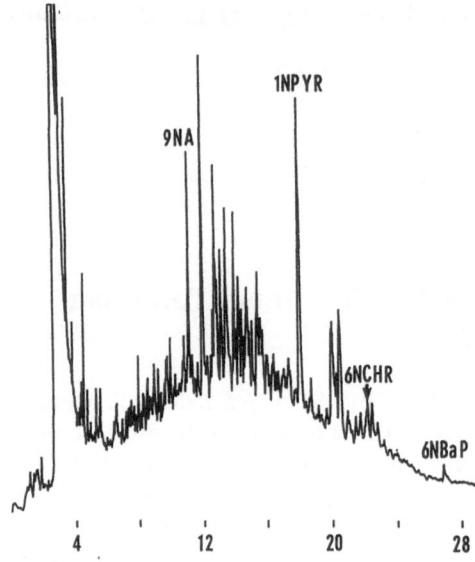

Fig. 3. GC (ECD) chromatogram of a Nitro-PAH mixture isolated from NBS Standard Reference Material 1650 (Diesel particulate matter)

Table 3. PAH and nitro-PAH concentrations in Standard Reference Material 1650 (diesel particulate matter) in comparison with certified values specified by NBS

PAHs	Concentrations [μg/g]		Recovery Rate [%]
	Values stated by NBS	Analyzed Value	
PHE[a]	71	52 ± 12[b]	73 ± 23
FLU*	51 ± 4	51 ± 9	99 ± 18
PYR*	48 ± 4	50 ± 8	103 ± 16
B(a)A*	6.5 ± 1.1	6.2 ± 2	95 ± 32
CHR/TRI	22	26 ± 5	118 ± 19
BbF	–	n.d.[c]	n.d.
BkF	2.1	n.d.	n.d.
BaP*	1.2 ± 0.3	0.8 ± 0.2	68 ± 25
I1,2,4-cdP	2.3	2.6 ± 0.4	113 ± 15
BghiP*	2.4 ± 0.6	2.5 ± 0.3	104 ± 12
Nitro-PAHs			
NA[d]	–	8.6 ± 1.5	
NPYR	19 ± 2	8.7 ± 1,2	46 ± 14
NCHR	–	2.2 ± 0.8	
NBaP	1.6	6.0 ± 1.9	

[a] Abbreviations, see Table 2;
[b] Standard deviation (n = 5);
[c] not detected;
[d] Abbreviations used: NA = 9-Nitroanthracene, NPYR = 1-Nitropyrene, NCHR = 6-Nitro-chrysene, NBaP = 6-Nitrobenz(a)pyrene;
* Certificated Values

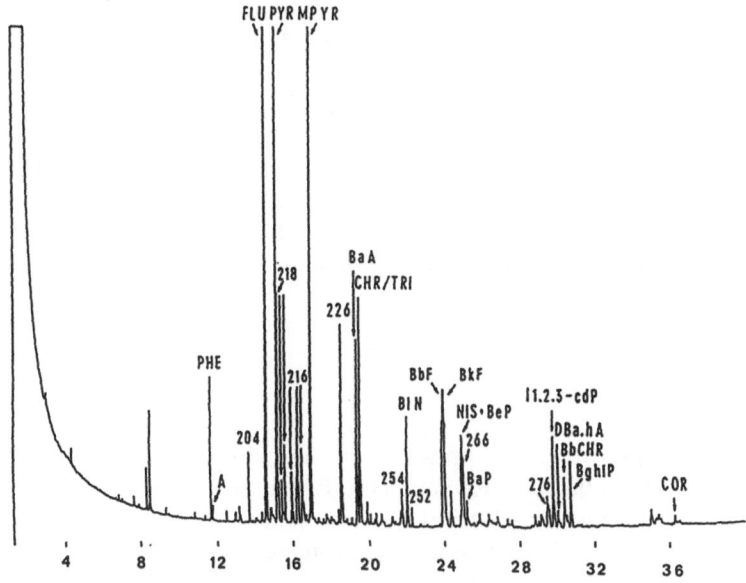

Fig. 4. GC (FID) chromatogram of a PAH mixture isolated from a fly ash (sampled during the combustion of a fuel oil on Teflon coated glass fiber filters)

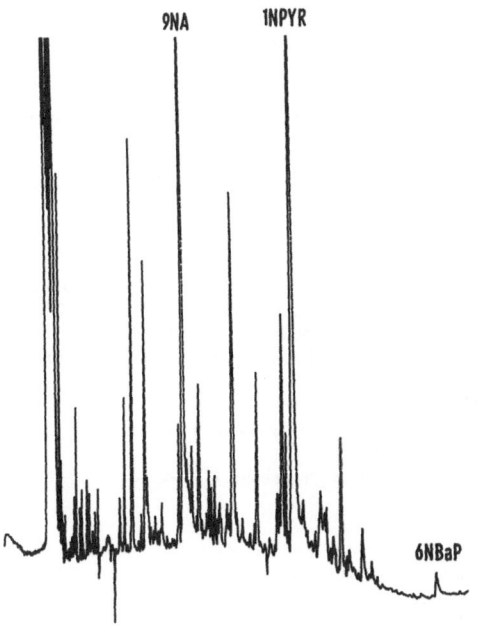

Fig. 5. GC (ECD) chromatogram of a Nitro-PAH mixture isolated from a fly ash (sampled during the combustion of a fuel oil on Teflon coated glass fiber filters)

the cleanup step described before. The column eluants were quantitatively analysed for PAHs and qualitatively for Nitro-PAHs by GC (FID) and GC (ECD), respectively.

Figures 4 and 5 present the gas chromatograms of the PAH and Nitro-PAH compounds found on the filters. The compositions of the GC (FID) and GC (ECD) chromatograms shown in these figures were verified by GC-MS. The numbers at the top of the peaks (Fig. 4) represent the masses of molecular ions of PAHs which could not be identified. The origin of the Nitro-PAHs has not been verified. An artifact-formation during the sampling procedure as described by Schuetzle [37] cannot be excluded.

Conclusion

The single step cleanup procedure presented in this paper is a simple, rapid and low cost method. It allows PAHs and Nitro-PAHs, and if necessary PCBs, extracted even from complex matrices (biological and environmental) by conventional Soxhlet-extraction to be isolated from accompanying substances and to be quantitated simultaneously.

Due to the good results obtained also with environmental and biological materials the cleanup procedure seems to be well suited for almost all of the materials stored in the Environmental Specimen Bank.

References

1. Lee ML, Novotny MV, Bartle KD (1976) Anal Chem 48: 1566
2. Grimmer G, Böhnke H, Glaser A (1977) Zentralbl Bakteriol Parasitenkd Infektionskr Hyg Abt I Orig Reihe B 164: 218
3. JARC Monographs on the Evaluation of the Carcinogenic Risk of Chemicals to Humans. a (1973) Vol 3, b (1983) Vol 32 International Agency for Research on Cancer, Lyon
4. Madsen ES, Nielsen PA, Pedersen JG (1982) Sci Total Environ 24: 13
5. Cooke M, Dennis AJ (eds) (1983) Polynuclear aromatic hydrocarbons: Formation, presence, and metabolism, Batelle Press, Columbus OH
6. Grimmer G, Jacob J, Naujock KW, Dettbarn G (1983) Anal Chem 55: 892–900
7. Lawrence JF (1987) Chromatographia 24: 45–50
8. Bosch LC, Chan TL, Duncan WP, Gibson TL, Jenson TE, Mermelstein R, Perry M, Poole CF, Rosenkranz HS, Ruehle PH, Schuetzle D, Tomkins BA, White CM (1985) In: White CM (ed) Nitrated polycyclic aromatic hydrocarbons. Dr. Alfred Huethig Verlag, Heidelberg
9. Karasek FW, Hutzinger D, Sate S (eds) (1985) Mass spectrometry in environmental sciences. Plenum, New York
10. May WE, Wise SA (1984) Anal Chem 56: 225–232
11. Grimmer G, Naujock KW, Schneider D (1982) Fresenius J Anal Chem 311: 475–484
12. Chaudhury DR, Bush B (1984) Anal Chem 53: 1351–1356
13. Stray H, Mano S, Mikalsen A, Oehme M (1984) J High Resolut Chromatogr Chromatogr Commun 7: 74–82
14. Das BS, Thomas GH (1978) Anal Chem 50: 967–973
15. Wise SA, Bowie SL, Chester SN, Cuthrell WF, May WE, Rebbert RE (1982) Cooke M, Dennis AJ, Fischer GL (eds) In: Polynuclear Aromatic Hydrocarbons: Physical and Biological Chemistry Batelle Press, Columbus OH, pp. 919–928
16. Fechner D, Seifert B (1978) Fresenius J Anal Chem 292: 199–202

17. Grimmer G, Böhnke H (1975) Cancer Lett 1: 75–84
18. Kunte H, Borneff J (1976) Z Wasser-Abwasser-Forsch 9: 35–38
19. Grimmer G, Hildebrandt A, Böhnke H (1975) Dtsch Lebensm Rundsch 71: 93–100
20. Schuetzle D, Lee FS-C, Prater TJ, Tajada SD (1981) Int J Environ Anal Chem 9: 93–144
21. Paputa-Peck MC, Marano RS, Schuetzle D, Riley TL, Hampton V, Prater TJ, Skewese LM, Jensen TW, Ruehle PH, Bosch LC, Cuncan WP (1983) Anal Chem 55: 1946–1954
22. Newton DL, Erickson MD, Torner KD, Pellizzari ED, Gentry P, Zweidinger RB (1982) Environ Sci Technol 16: 206–213
23. Nielsen T (1983) Anal Chem 55: 286–291
24. D'Agostino PA, Narine DR, Mc Carry DE, Quilliam MA (1982) Cooke M, Dennis AJ (eds) In: Polynuclear Aromatic Hydrocarbons: Formation, Metabolism and Formation, Batelle Press, Columbus OH, pp. 603–613
25. Campbell RM, Lee ML (1984) Anal Chem 56: 1026–1030
26. Zielinski B, Arey J, Atkinson R, Winer AM (1989) Atmos Environ 23: 223–229
27. Ciccioli P, Cecinato A, Brancaleoni E, Draisci R, Liberti A (1989) Aer Sci Techn 10: 296–310
28. Onuska FJ, Tery KA (1989) J High Resolut Chromatogr Chromatogr Commun 12: 362–367
29. Schilhabel J, Levsen K (1989) Fresenius J Anal Chem 333: 800–805
30. Mac Crehan WA, May WE, Yang SD, Benner Jr BA (1988) Anal Chem 60: 194–199
31. Bayona JM, Markides KE, Lee ML (1988) Environ Sci Technol 22: 1440–1447
32. Roerden O, Reisinger K, Leymann W, Frischkorn CBG (1989) Fresenius J Anal Chem 334: 413–417
33. Kloster G, Niehaus R (long term sampling of air for the determination of PAHS) (to be published) .
34. Uhlig E, Doll F, Hackfort H, Zajc A (1989) VDI Bericht 765 447456
35. Wolf J (1988) Patentschrift DE 3702 89 Cl Patentamt München
36. Reisinger K, Leymann W (to be published)
37. Schuetzle D (1983) Environ Health Perspective 47: 65–80

5.3 Analysis of Chlorinated Hydrocarbons (CHC) in Environmental Samples

K. Oxynos, H. Schmitzer, H.W. Dürbeck, and A. Kettrup

Chlorinated Hydrocarbons (abbreviated CHC* in this standard operation procedure) play an important role as environmental contaminants. Predominantly in the past they have been intensively used worldwide as effective insecticides (DDT, HCH and related compounds) or as transformer oils (PCBs). However, as a consequence of the hazardous side-effects observed mainly in animals, birds, and primates the application of those pesticides was strictly banned in the early seventies and the production of PCBs was drastically reduced. Nevertheless, due to the low degradability of CHCs and their high accumulation-factors in fatty tissues they are still distributed all over the world; hence they are most suitable as stable indicators of the status and the development of the environment, particularly in the context of an Environmental Specimen Bank.

1 Introduction

1.1 Objective

The objective of this Standard Operation Procedure (SOP) is not only to describe a uniform method for the analytical determination of chlorinated hydrocarbons applicable to all matrices of the environmental specimen bank (ESB) collected so far, but also to emphasize some critical parameters and operational steps of the methodology.

1.2 Range of Application

All ESB materials selected so far – including human specimens – can be analyzed by the method presented here.

1.3 Schematic Description

The individual steps of the entire procedure are schematically shown in the scheme below. In Fig. 1, a detailed description of the method is represented by the flow-diagram, where the used symbolism [1] coincidently gives an insight into the applied physico-chemical principles.

---

*CHC = α-, β- and γ-HCH, HCB, DDE, Aldrine, Dieldrine, Heptachlor (HC), Heptachlorepoxide (HE) and PCBs Clophen (A-60).

Specimen Banking
Rossbach/Schladot/Ostapczuk (Eds.)
© Springer-Verlag Berlin Heidelberg 1992

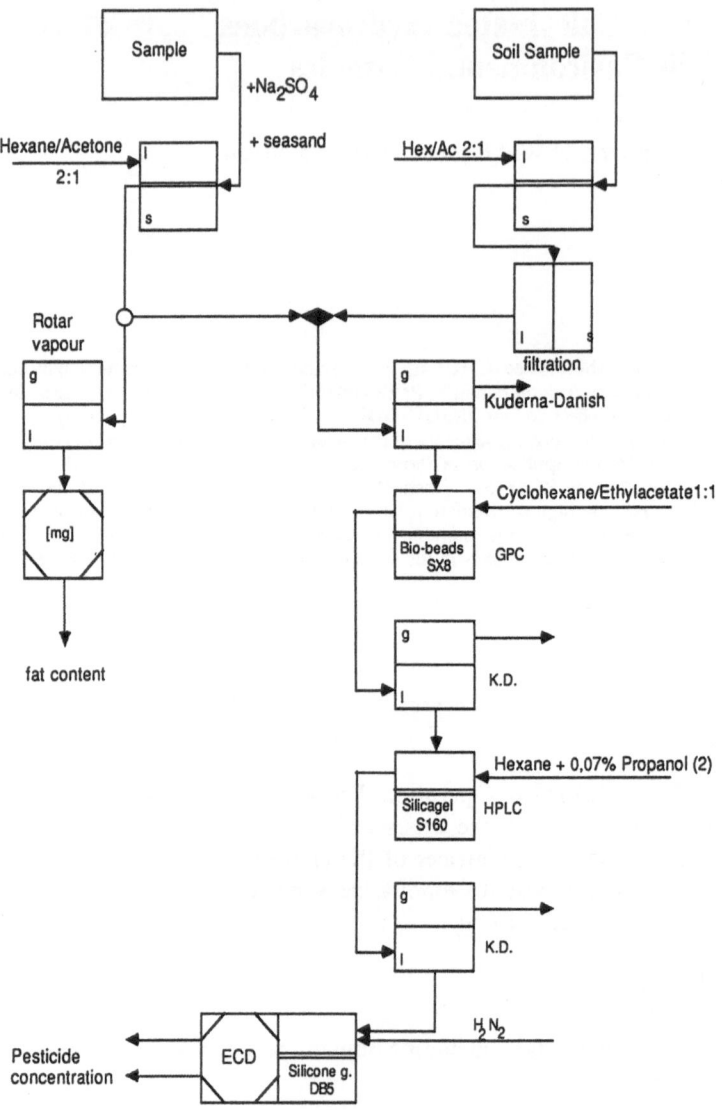

Fig. 1. Detailed scheme of the analytical procedure (g = gas phase; l = liquid phase; s = solid phase)

Sample scheme

- Mixing with anhydrous sodium sulfate/sea sand (2:1)
- Column extraction with *n*-hexane/acetone (2:1)
- Subsample: gravimetric determination of fat
- Gel permeation chromatography

- Evaporation at the Kuderna-Danish-apparatus
- High-performance liquid chromatography (separation of fat)
- Evaporation by means of the Kuderna-Danish-apparatus
- Capillary gas chromatography

2 Materials and Experimental Prerequisites

2.1 Chemicals/Solvents/Accessories

The quality of the solvents and chemicals is at least of analytical grade and has to be individually checked for its specific suitability. Moreover, these materials as well as any supporting accessories are subjected to permanent internal laboratory control (see Sect. 5) to guarantee unchanged long-term purity and quality.

2.2 Standards and Calibration Mixtures

Calibration solutions and reference mixtures which are prepared from commercial standards are also subjected to the internal laboratory control. Their concentrations are usually in the ppb-level, ranging typically from 10 to 100 pg/μl.

2.3 Glassware

Each individual part of the necessary glassware is exclusively used only in CHC-trace analyses. Only brand-new articles are acceptable which are conditioned with a CHC standard mixture before primary use.

2.3.1 Precleaning of glassware

Initially, the following steps have to be strictly performed:

- Precleaning in a dishwasher with deionized water,
- drying at 150 °C in a cabinet dryer and intermediate storage under "clean" conditions until utilization.

2.3.2 Additional cleaning steps directly before use:

- rinsing 3 times with acetone (analytical grade),
- rinsing 3 times with n-hexane (analytical grade),
- drying in a hot air stream.

(In general, the application of laboratory grease is not permitted; the connection of the ground joint equipment is performed by means of precleaned Teflon packing rings).

2.4 Equipment

2.4.1 Specimen Storage and Intermediate Storage of Extracts

- Cryovessels: liquid nitrogen (LN$_2$), from -130 to $-160\,°C$;
- deep freezers: $-20\,°C$.

2.4.2 Homogenization

- Rotary mill (Retsch Co.), sieve diameter 2 mm; 4000 rpm;
- auxiliary tools manufactured from stainless steel, Teflon or porcelain.

2.4.3 Extraction

- Glass columns; length (l) = 50 cm, internal diameter (i.d.) = 2 cm with glass or Teflon stopcocks;
- muffel furnace, heatable up to 800 °C.

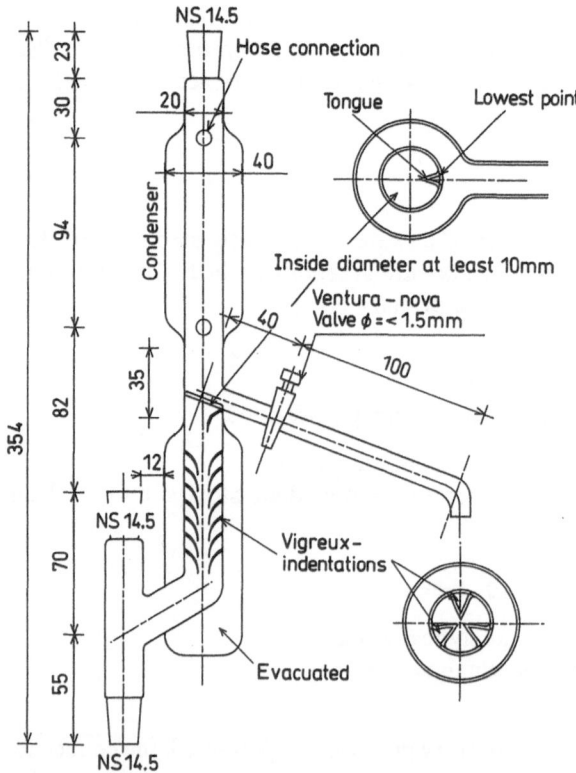

Fig. 2. Schematic drawing of the modified Kuderna-Danish apparatus

2.4.4 Evaporation of Extracts

– Rotary evaporator;
– modified Kuderna-Danish-apparatus (Fig. 2);
– vacuum Kuderna-Danish-apparatus attached to a water jet vacuum system, equipped with a constant pressure controller.

2.5 Cleanup of Crude Extracts

2.5.1 Gel Permeation Chromatography

– Gel permeation system (GPC) totally automated, including sample application and collection of the CHC-fraction; column LC-61123 (variable length; i.d. = 127 mm); stationary phase: BIO-Beads-SX-8 (Biorad Co.); mobile phase: cyclohexane/ethylacetate (1 : 1), 2 ml/min.
– GPC as described before, but using benzene as mobile phase.

2.5.2 High Performance Liquid Chromatography (HPLC)

– High performance liquid chromatography system, totally automated as described in 2.5.1 including time-controlled pre- and reconditioning (cf. 3.3.2); column: $l = 24$ cm, i.d. = 0, 8 mm; stationary phase: Lichrosorb 5 um, Silicagel SI 60, mobile phase: 0.1% 2-propanol in hexane, 4 ml/min,
– benchtop centrifuge (3000 rpm).

2.6 Gas Chromatography

– Capillary gas chromatograph equipped with electron capture detector (ECD) and auto-sampler; fused silica capillary column coated with OV-101 or DB-5, respectively; both as bonded phase; $l = 30$ m, i.d. = 0.32 mm;
– carrier gas: H_2 at 2 ml/min;
– temperature program (for example): 60 °C to 130 °C with 30 °C/min, 3 min isothermal, then to 180 °C with 5 °C/min and further with 8 °C/min to 280 °C;
– sample amount injected: 1 μl, splitless;
– injector temperature: 280 °C.
– Instrumental equipment (for example): Hewlett Packard 5890A; Siemens-Sichromat II or Carlo Erba Mega-Series.
– Data system: laboratory data system LAS (Hewlett-Packard).

3 Experimental Performance

3.1 Specimen Storage and Homogenization

– Before analysis, homogenized samples are constantly kept at LN_2 temperature.

– Unhomogenized specimens are homogenized and subsampled in the deep
  frozen state using the precooled (LN$_2$) rotary mill (cf. 2.4.2). In every case
  representative subsamples have to be taken from the homogenized material.

3.2 Analytical Procedure

3.2.1 Primary Extraction

In a mass ratio of 1:4 the sample is added to anhydrous sodium sulfate/seasand
(2:1; w:w) and mixed in a porcelain mortar to form a free-flowing product,
which is then extracted with 300 ml of an *n*-hexane/acetone mixture (2:1; v:v) in
an extraction column (cf. 2.4.3).

3.2.1.1 Precautions to be Noted

– The direct contact of the unmixed sample with the surface of the mortar has to
  be avoided.
– When filling the extraction column, a homogenious distribution of the
  extraction mixture has to be ensured.
– The extraction time should be at least 3 hours.
– The volume of the eluate should be at least 280 ml.
– Irregularities in the extraction procedure have to be notified in the analytical
  document.

3.2.2 Application of the Internal Standard (istd)

Both internal standards (pentachlorobenzene, pcb, and decachlorobiphenyl,
dcb) are added to the base extracts. Typically applied amounts are pcb/dcb:
10/20 ng or 20/40 ng.

3.2.3 Aliquotation of Base Extracts

Aliquotation of the base extracts by gravimetry is only performed if a determination of fat is intended (cf. Sect. 3.2.4). Due to the limited fat separation capacity
of the GPC and HPLC systems aliquotation is mandatory in case of samples
with a high fat content. Further aliquotation is necessary if the capacity is still
being exceeded.

3.2.4 Gravimetric Determination of Fat

– An exactly defined part of the base extract (cf. Sect. 3.2.3) is evaporated in a
  weighed flask up to a constant weight by means of a rotary evaporator. The
  fat content is directly calculated from the weight of the dry residue.
– In general, the analytically obtained CHC-data are *not* related to the fat
  content.

3.2.5 Evaporation of the Base Extracts

– The evaporation of the crude extracts to a volume of approximately 5 ml is performed by a modified Kuderna-Danish-apparatus (cf. Fig. 2).
– Due to the high volatility of the low molecular weight of CHCs, extreme care has to be taken if a concentration with a volume of less than 5 ml is intended.
– In any case, evaporation to dryness has to be strictly avoided.

3.2.6 Freezing and Centrifugation

For plant materials (e.g. spruce needles) a separation of the precipitating solid matter by means of freezing and centrifuging of the 5 ml extracts (cf. Sect. 3.2.5) is necessary. The liquid phase is decanted and then carefully concentrated to approximately 2 ml.

3.2.7 Desulfurization of Sewage Sludge Extracts

– 2 ml of bidistilled water and 200 mg Cu powder are added and carefully mixed (ultrasonic bath) with the concentrated extract of Sect. 3.2.5, and refluxed for 2 hours at 70 °C.
– After filtration the aqueous phase is separated. The organic layer is dried with Na_2SO_4 and concentrated to about 2 ml. For further treatment see Sect. 3.2.8.

3.2.8 Exchange of Solvents

– For the cleanup by GPC (cf. Sect. 3.3.1) an exchange of the solvent is mandatory. This exchange is accomplished with the Kuderna-Danish apparatus by adding 10 ml of the *new* solvent and concentrating to approximately 2 ml. This procedure is repeated at least 3 times.
– *Nonplant* materials are made up in 2 ml of ethylacetate.
– *Plant* matrices are transfered to about 2 ml of benzene.

Notice !!!: N-hexane has to be removed *completely* from the base extracts since even very small amounts of this solvent significantly influence the swelling behaviour of the GPC-columns.

3.3 Cleanup Procedure of the Base Extracts

3.3.1 By gel permeation chromatography (GPC)

– A pre- and reconditioning of the column is unnecessary.
– Extracts of *nonplant* matrices are chromatographed with cyclohexane/ethyl-acetate (cf. Sect. 2.5.1) collecting the specific CHC-fractions.
– The extracts of *plant* materials are passed through the benzene GPC-column for separation of the CHC-fraction.
– The typical CHC-fractions are determined by GPC of a CHC standard mixture under the same experimental conditions.
– The fractions are evaporated to about 2 ml (cf. Sect. 3.2.5) and transfered into n-hexane (cf. Sect. 3.2.8).

3.3.2 By High Performance Liquid Chromatography (HPLC)

3.3.2.1 Pre- and Reconditioning of the System

The automatic and time regulated pre- and reconditioning of the HPLC-system is performed after a column change or immediately after the cleanup of each individual sample (cf. Sect. 3.3.2.2) by:

- back-flush with 10 ml toluene
- back-flush with 40 ml n-hexane/0.1% 2-propanol
- rinsing with 40 ml n-hexane/0.1% 2-propanol.

3.3.2.2 Chromatography

- The concentrated CHC-fractions (1.5–2 ml hexane solution of 3.3.1) are further cleaned up by HPLC – collection of the specific CHC-fraction, previously determined by HPLC of a CHC standard mixture under the same experimental conditions.
- The eluate is concentrated to about 2 ml (cf. Sect. 3.2.5).

3.3.3 Notes on the Cleanup Procedures

- If required the cleanup steps (Sects. 3.3.1 and 3.3.2) have to be repeated.
- Suspended matter (prior to the GPC-procedure) as well as precipitating compounds (e.g. in case of the solvent exchange performed in Sect. 3.2.8) have to be removed.

3.4 Determination by Gas Chromatography (GC)

- In general, perfect gas chromatographic conditions with respect to parameter settings and column separation efficiency have to be always ensured. Optimization of these conditions (cf. Sect. 2.6) is carried out by analyzing suitable CHC standard mixtures.
- For quantification, a calibration with relevant concentrations (pg/μl) of this standard mixture has to be conducted at least every day.

3.4.1 Collection and Processing of Chromatographic Raw-Data

- Chromatographic raw-data of *one* sample are gained by at least 3 repetitive injections on 2 capillary columns of different polarity (OV-101, DB-5) resulting in a total number of 6 chromatograms.
- For the calibration procedure 3 relevant chromatograms of CHC standard mixtures for each column have to be available.
- Qualitative identification and validation is carried out by spiking the analytical sample with a reference mixture followed by computer-aided comparison of the obtained chromatograms.
- For safety reasons, after 3 sample determinations always one CHC standard mixture is analyzed.
- The resulting chromatograms are archived as analogous chromatograms (recorder output) and additionally stored as raw-data files (data system).

3.4.2 Quantitative Evaluation

– The quantitative evaluation is performed by the standard addition method using peak heights as reference signals.
– The calculation is based on the following equation:

$$Ci = \frac{IS \cdot EIS \cdot HI \cdot CCi}{SA \cdot HIS \cdot EI \cdot CCis} \quad [ng/g]$$

where

Ci	:	unknown concentration [ng/g] of compound i
IS	:	amount [ng] of internal standard in the sample
SA	:	amount of weighed sample [g]
EIS	:	peak height [mm] of the internal standard in the calibration chromatogram
HIS	:	peak height [mm] of the internal standard in the sample chromatogram
EI	:	Peak height [mm] of compound i in the calibration chromatogram
HI	:	Peak height [in mm] of compound i in the sample
CCi	:	amount of compound i [ng] in the calibration mixture
CCis	:	amount of int. standard [ng] in the calibration mixture

– additional specific corrections are not applied.
– By means of the first istd (pentachlorobenzene) the following CHCs are calculated: HCB, α-, β-, und τ-HCH, HC, HE, aldrine, DDE und dieldrine.
– The calculation of the PCB concentration (as unmetabolized Clophen A-60) is carried out using the second internal standard (decachlorobiphenyl).

3.5 Recovery Rate

– The recovery rates of all CHCs are determined by analyzing CHC standard mixtures and calculated by means of the internal standard method. The obtained results for both the individual cleanup procedures (GPC and HPLC) and the overall methodology have to be in the 90% range without any indication of time-dependent trends or losses.

3.6 Precision

The achieved precision – expressed as deviation (in %) from the mean – is determined as repeatability according to ISO 3354-1977, 2.84 from at least 5 repetitive CHC analyses covering a time span of 10 days. Typical values in the relevant concentration range (ng/g) should not exceed the 5–15% level.

3.7 Critical Limiting Value (Lc); Limit of Detection (Ld) and limit of Determination (Lq) [2]

– The data listed in Table 1 (μg/kg) are determined by the evaluation of analytical results obtained from blanks and spiked blanks.

Table 1. Typical results for Lc, Ld, Lq and Lq*, data in $\mu g/kg$

	HCB	αHCH	βHCH	τHCH	HC	Ald	HE	DDE	Diel	PCB
Lc	0.05	0.2	0.3	0.3	0.2	0.2	0.2	0.2	0.3	1.5
Ld	0.1	0.4	0.6	0.6	0.5	0.4	0.5	0.4	0.6	3.0
Lq	0.5	1.1	1.7	1.8	1.4	1.1	1.5	1.2	2.0	8.5
Lq[a]	0.5	2.0	2.0	2.0	2.0	2.0	2.0	2.0	2.0	10

[a] Average data

- The calculation is performed by the "internal standard method" assuming the weight of the sample to be 1 g.
- It should be noted that Lc, Ld and Lq are closely related to a number of individual parameters (matrix effects). In principle, these data should thus be quoted only for a particular chromatogram of a specific sample.

4 Results

Reported CHC concentrations of *one* sample are always mean values of at least 6 repetitive injections (see Sect. 3.4.1).

5 Internal Laboratory Control

- The reliability of the data obtained demands a continuous control of all operational steps with respect to their reproducibility and repeatability. This protocol includes the repetitive check of all chemicals, gases and auxiliary tools used.
- Moreover, standard solutions (mixtures and individual compounds) are regularly controlled with respect to their qualitative and quantitative composition.
- Within any analytical series a blank sample and a spiked blank sample are additionally analyzed.
- Calibration data for quantitative analysis are redetermined daily and compared with previous results.
- In case of deteriorations spiked blank samples are analyzed to check the individual operational steps. In addition, the recovery rates are reexamined to trace for systematic errors.
- The complete raw-data files are stored long-term for repeated evaluations at any time.

6 Conclusion

An essential feature of the described method is the cleanup procedure which is applicable to a great number of different matrix types. Fundamental part of this procedure is the tailor-made combination of GPC and HPLC. The CHC containing fractions which have to be redetermined only in the case of a necessary column change are independent of the instrumental equipment used. They are governed only by the behaviour of the CHCs in the exactly defined chromatographic system (stationary phase and solvent). After a column change differences in the retention time normally do not exceed the ± 5% level. The stability of *one* GPC-column is at least 6 months while the HPLC-system may be used for more than 1 year without any column change.

Based on the considerably automated performance, the entire analytical procedure is more or less protected against individual errors, thus providing the necessary stability for standardization, which is essential for obtaining comparable results in the analysis of different environmental specimens. Hence, not only reliable data for matrix related long-term trend assessments will be available, but also direct correlations between various environmental specimen types seem to be feasable.

References

1. Rohleder H, Gorbach SG (1979) Z Anal Chem 295: 342
2. Beyermann K (1982) Organische Spurenanalyse. Georg Thieme Verlag, Stuttgart p 20 (Monographien über "Analytische Chemie für die Praxis)

6 Inorganic Analytical Approaches

6.1 Nuclear Analytical Methods in Environmental Specimen Banking

Susan F. Stone and Rolf Zeisler

Several nuclear analytical methods have been applied to determine elemental concentrations in samples from the National Biomonitoring Specimen Bank. Various combinations of neutron activation analysis (NAA), prompt gamma activation analysis (PGAA), and X-ray fluorescence (XRF) used to obtain information on the inorganic constituents in four types of samples, human livers, marine bivalves, fish livers, and sediments, are described. Concentrations of at least 20–30 elements from a small test portion (~ 1 g wet weight) can be obtained by combining these methods. The dynamic ranges, sensitivities and multielement capabilities of the described methods are shown to provide the high quality data needed in a banking program.

Introduction

Of the numerous inorganic analytical methods available, nuclear methods are particularly well-suited for obtaining substantial information on the elemental composition of various environmental and biological samples [1], and have been successfully applied to the characterization of samples from the National Biomonitoring Specimen Bank (NBSB). The NBSB program has already been described in a previous chapter of this book and elsewhere [2]. Samples stored in the NBSB are often unique and are limited in quantity, only allowing about 20 tests per banked sample. Hence, particular attention has been given to the planning of the analytical process to extract the maximum amount of information from a test portion and to assure the quality of the analytical data. This chapter describes how and why nuclear analytical methods are used in the NBSB program. Although the main emphasis will be on neutron activation analysis (NAA), another nuclear method, prompt gamma activation analysis (PGAA) and a related technique, though not strictly nuclear, x-ray fluorescence (XRF) are useful complements for sequential analysis of complex environmental samples, such as those from the environmental specimen bank.

Even though all stable elements (and some radioactive elements) may be contained in the banked samples, knowledge about the biochemical and physiological role of elements is limited to relatively few. These include the major and minor constituents in biological matrices, mineral elements, essential trace elements, and a small number of elements that have known adverse effects in biological and environmental systems at trace levels. Commonly, not all of these better known elements are determined even in several aliquots of a given sample,

Specimen Banking
Rossbach/Schladot/Ostapczuk (Eds.)
© Springer-Verlag Berlin Heidelberg 1992

let alone determining elements that are rarely or never considered. Consequently, the concentrations of the majority of the elements are little-known and are not considered in evaluations of biological and environmental samples. However, increased knowledge of elements in biological and environmental samples is needed to assess health issues and to study biological pathways and environmental source contributions.

One of the many advantages of nuclear techniques is their ability to simultaneously and sequentially determine many elements in a single sample. Since these methods depend on nuclear properties, analyses are independent of the chemical state, and so they are able to assay the total amount of element (instead of by oxidation state, etc.), and avoid problems of chemical interferences. Different matrices generally do not affect the analyses, and by using nondestructive methods (no sample dissolution), there is little or no sample preparation, so that the effects of blank, losses, and chemical yield are usually negligible. The nuclear techniques have broad dynamic ranges and, depending on the element, concentrations from pg/g to percent levels can be quantified. The sensitivity for many elements obtained by NAA is particularly good; it is possible to quantify less than pg amounts of an element in a sample, with the added advantage of not diluting the sample. Dissolution is unnecessary prior to analysis with this method, but is required in many other techniques. Concentrations of at least 20–30 elements from a small sample ($\sim$ 1 g, wet weight) can be obtained by combining several nondestructive nuclear methods.

Nuclear Methods for Instrumental Assay

Neutron Activation Analysis

For the method of NAA, nuclides of the stable elements are transformed to radionuclides by irradiation of the sample with neutrons. Nuclei of different isotopes within a sample have various probabilities of capturing the neutrons. Following neutron capture, many product nuclei decay by particle and/or gamma ray emission. Detection of these decay products is utilized for quantification of the elements.

The activity of a particular nuclide in an irradiated sample at a given time is proportional to a number of factors, including the amount of the isotope, probability of neutron capture (known as the cross section), neutron fluence, and the product of the decay constant of the nuclide to be measured and the time of decay. If samples and comparator multielement standards are treated identically (including irradiation position and time, detector counting geometry and efficiency), and if they have the same isotopic abundances, the amount of an element in a sample, m_{unk}, can be calculated by the comparator method using the equation:

$$m_{unk} = \frac{A_{0(unk)}}{A_{0(std)}} m_{std}$$

where A_0 is the activity at the end of the irradiation.

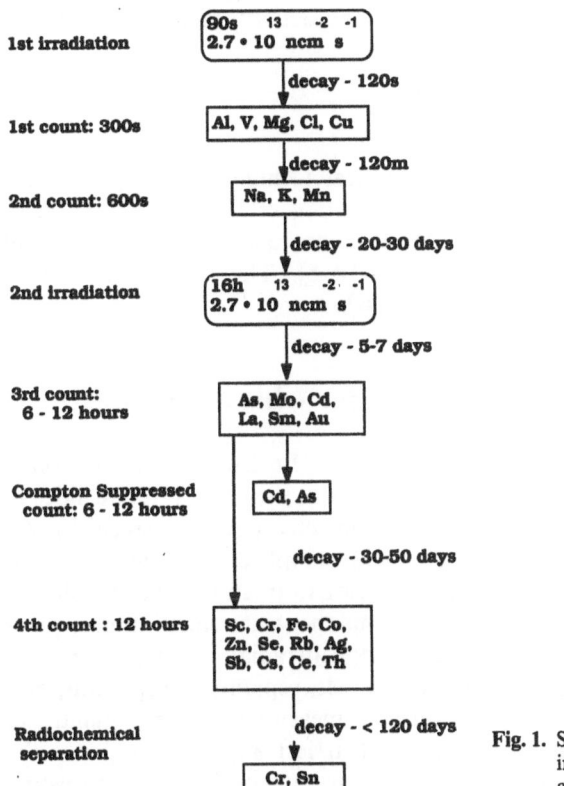

Fig. 1. Schematic flow chart for a typical instrumental neutron activation analysis

A diagram of a typical series of irradiations and counting cycles needed to obtain concentrations for 28 elements in a liver sample by instrumental neutron activation analysis (INAA) is shown in Fig. 1. This figure illustrates the suites of nuclides with similar half lives obtained simultaneously with each irradiation/counting step. The selection of various irradiation and decay times emphasizes the detection of one group of nuclides over others.

Prompt Gamma Activation Analysis

A complementary method to NAA is neutron-capture prompt gamma-ray activation analysis (PGAA). In PGAA, the emitted gamma rays are detected while the sample is being irradiated with neutrons. Following neutron capture, the excitation energy is released by emission of one or more "prompt" gamma rays. PGAA has widely varying sensitivities for all elements [3], which depend on the capture cross sections and gamma ray yield for a specific gamma ray. PGAA is particularly useful in biological samples to quantify light matrix elements, as well as several major and trace constituents. Quantification in PGAA is accomplished by the comparator method, similar to the NAA example above, or by the use of a flux monitor, where count rates of elements at certain fluence rates are known and are related to the activity of a single element monitor.

X-ray Fluorescence

X-ray fluorescence (XRF) is another analytical technique employed in the analysis of NBSB samples. A particular application of XRF, using the back-scatter with fundamental parameter (XRF-BFP) procedure, is especially useful for biological and environmental matrices [4, 5]. The specific conditions followed for these analyses can be found in [6].

Nuclear Methods with Radiochemistry for Ultimate Sensitivity

To achieve the best sensitivity in NAA for low level trace elements, it is often necessary to separate the element(s) of interest from the other radionuclides in the sample. In these cases, radiochemical neutron activation analysis (RNAA) is necessary. Most of the problems associated with sample dissolution are elimin-ated in RNAA since the sample is activated prior to treatment, i.e. dissolution can be performed in an essentially blank-free manner since chemical manipula-tion of the sample is done after the activation (irradiation) step. The non-activated elements in reagents, dissolution vessels, separation apparatus, etc., cannot affect the amount of radioactive nuclides measured. In fact, a quantity of the stable (unirradiated) element(s) of interest (carrier) is typically added to the sample prior to chemical manipulation to minimize the problems of processing dilute solutions.

Many different types of radiochemical separations can be performed, from precipitation methods to extractions and ion exchange. Two examples of RNAA procedures that have been used in the NBSB program are liquid–liquid extractions for Cr and Sn [7], and the use of an inorganic ion exchange column for a group separation for Cr, As, Se, Mo, Ag, Sn, and Sb [8].

Data Quality

An essential point in any chemical characterization is that there should be a means to assure the quality of the data. Often, data can be confirmed by replicate analyses and independent determination of sample constituents by different analytical techniques. With the banked samples, however, replicate analyses by numerous techniques are not practical on account of the limited sample amounts. Multielement analysis of inorganic constituents is used in place of replicate analyses by one technique to eliminate the occasional experi-mental blunder. Several elements can be determined by more than one method, and in the case of NAA, an element can sometimes be determined in more than one irradiation/counting cycle. Also, systematic errors that could arise from the use of a single technique can often be detected by observing results from the suites of elements with nuclides with similar half lives. For instance, if concentrations obtained from one NAA sequence are consistently low compared to a previous analysis of the same sample, a weight basis problem might be detected. Also, since the nuclear methods used here (with the exception of radiochemistry steps) are nondestructive, most steps can be repeated if there is some question about a specific data point, or, in specific instances, a sample can be reanalyzed under more optimum conditions to enhance sensitivity.

Another important part of checking data quality in the NBSB program has been the analysis of certified reference materials along with the samples. Several types of certified reference materials are included in each set of analyses. For example, the results obtained for zinc on NIST Standard Reference Material (SRM) 1577 and 1577a, Bovine Liver, analyzed along with the human liver samples, are summarized in Fig. 2. Since the zinc concentrations differ slightly in the two materials, the results are normalized to the certified value which is set equal to one. Each value is shown with its uncertainty (1σ) due to the counting statistics of samples and standards. The observed relative standard deviation (1σ) of the data is 2.5%. During the ten year period of inorganic analyses for the NBSB, major changes in nuclear instrumentation have occurred which provide improved spectral resolution as well as more sophisticated peak analysis software. These improvements have not led to changes in the measured concentrations. Also, the fact that different personnel have had principal responsibility

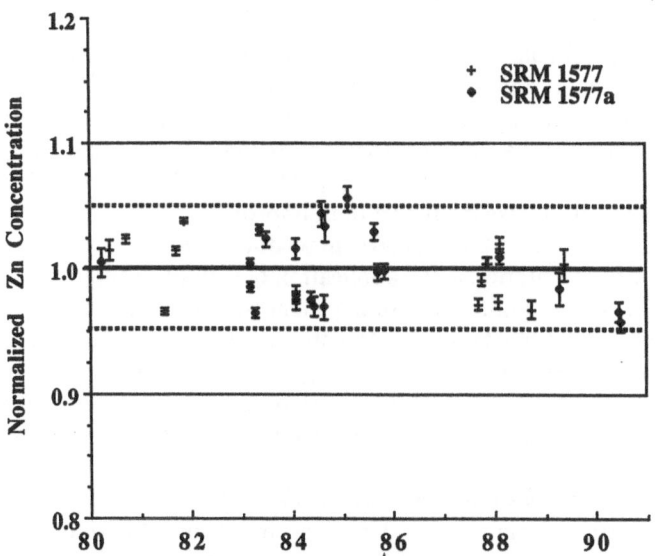

Fig. 2. Control chart for Zinc in SRM Bovine Liver from 1980–1990
Normalized Zn concentrations are Zn (observed)/Zn (certified) for the two SRMs, SRM 1577 and 1577a. Dashed lines represent 2 (± 5%) observed relative standard deviation. Solid lines represent bounds for certified values (± 10%)

for individual measurements throughout the studies and that, on occasion, a different reactor has been used for irradiations has had no significant influence on the results. A larger scatter in the data was observed during 1984–1985 when shorter and earlier instrumental counting was done for the liver samples to allow for a subsequent radiochemical separation, but the results are still within the uncertainties of the certified values. This illustrates the consistency of the NAA technique.

Application of Nuclear Methods in the NBSB Program

Four types of samples that have been analyzed by nuclear methods in the NBSB program will be described: 1) human livers, from the joint National Institute of Standards and Technology (NIST)/Environmental Protection Agency (EPA) sponsored program; 2) marine bivalves, initially from EPA and more recently from the National Oceanic and Atmospheric Administration (NOAA) sponsored National Status and Trends (NS&T) program; 3) fish livers; and 4) marine sediment samples, the latter samples also from the NOAA NS&T program.

Human Livers

The inorganic scheme applied to the human livers has been modified since the beginning of the NBSB program. It has evolved from the use of several analytical methods [9], i.e. NAA, atomic absorption spectroscopy (AAS), isotope dilution mass spectrometry (IDMS), and anodic stripping voltammetry (ASV), obtaining elemental concentrations by relying almost totally on NAA [10]. In addition to the data obtained by NAA, a collaboration in recent years with the Environmental Specimen Bank of the Federal Republic of Germany has provided data determined by voltammetry for Pb and Ni, and Hg by cold vapor AAS, as well as comparative results for Cu, Cd, and Zn by voltammetry.

The changes in the analysis scheme reflect both improvements in instrumentation and the application of radiochemistry procedures. One improvement involving analytical instrumentation is the application of Compton suppression [11] to improve the detection limits for several elements, especially for As, Cd, and Cr. For the assay of short-lived nuclides, loss-free counting has been utilized to handle count rates up to 70,000 cps with a typical detector resolution of 1.95 keV FWHM at 1332 keV. The radiochemical procedures applied to the liver samples have been expanded from single element determinations for Sn and Cr to a group separation [8]. This has been done to obtain better sensitivity for several low level trace elements simultaneously. In some cases, information has been obtained from both instrumental and radiochemical neutron activation analysis (INAA and RNAA), with the RNAA results having lower uncertainties.

A summary of the concentrations for 28 elements obtained by NAA in the NBSB human liver analyses is given in Fig. 3. The data in the figure represent results for up to 96 individual human liver samples. As has been discussed

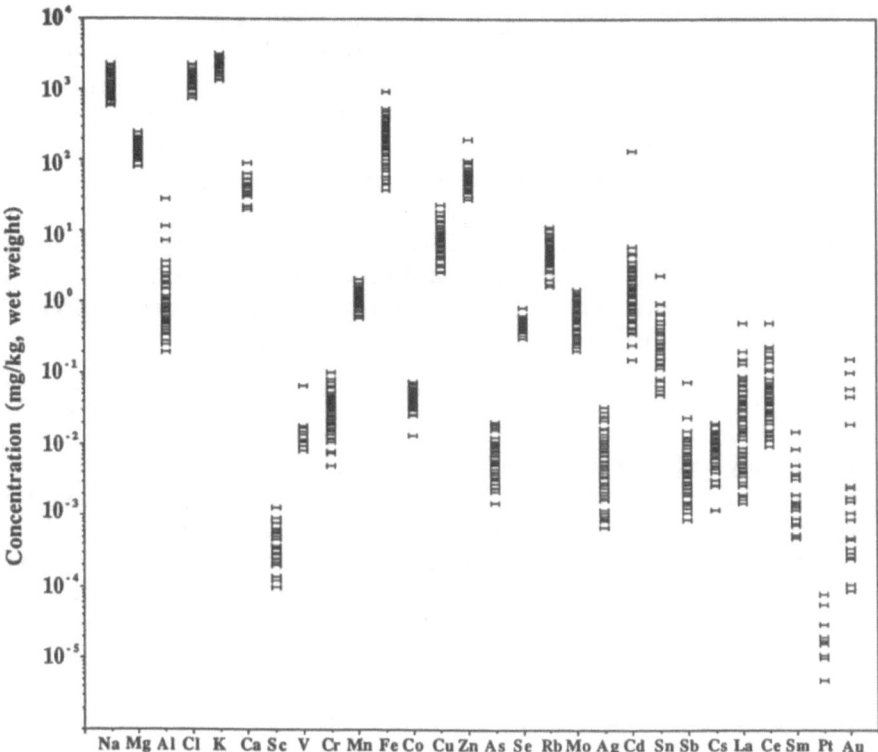

Fig. 3. Concentrations for 28 selected elements (mg/kg, wet weight) in 96 human liver samples

previously [12], it is significant that a very narrow range of values has been obtained for many essential trace elements (e.g. Na, Se, Zn) while the values for pollutant elements (e.g. Al, As, Cd, Ag, Sn) show greater variability. At the time the first set of results was obtained, the values found for many of the pollutant trace elements were at concentration levels much lower than expected, and often below previously reported data [13]. This appears to result from carefully minimizing contamination of the samples in this work. Contamination has been carefully controlled during sample collection, and this control has continued throughout storage prior to analysis and sample preparation, as well as throughout the entire analytical procedure. Many more researchers have now recognized the importance of early contamination control and this has been confirmed by a recent literature compilation [14] showing many of the trace element concentrations in human livers at similarly low levels.

Work on minimizing contamination still needs to continue, as illustrated by the levels of Cr obtained from the human liver homogenates. The levels (100 mg/kg and below) obtained for the liver homogenates are low compared to previous values; however, recent data suggest that even homogenization with

Teflon disk mills is contaminating the liver homogenates with irregular amounts of Cr, which are now known to be present at trace levels in the Teflon. The actual levels of Cr in human livers are probably less than 20 mg/kg [15], so most of the values greater than this are most likely due to the Cr contribution from the homogenization process.

The large number of elemental concentrations determined in the human liver samples add much information to the literature in terms of baseline data. Few studies have such well-characterized liver samples with respect to their inorganic composition. In addition, these liver samples have been characterized with respect to their content of polychlorinated biphenyls and chlorinated pesticide residues [16].

Bivalves

A combination of nuclear techniques has been used for the design of the analysis scheme for the marine bivalves [6]. Nondestructive techniques are applied in sequence to the same subsample to obtain concentrations for more than 40 elements before the sample is consumed in a final radiochemical assay. This procedure involves, in sequential order, XRF-BFP, PGAA, INAA, and finally RNAA for the measurement of Sn. This scheme provides a high degree of internal quality control, even greater than those used for the human livers, since there are many elements that are determined independently. In addition, the series of analytical techniques can be applied in the above-mentioned sequence to a single 250 mg subsample.

In applying the procedure to bivalve samples, greatly differing inorganic concentrations are found. Concentrations for up to 45 elements in bivalve samples from the NOAA NS&T Mussel Watch program [17] are determined, and these concentrations range from percent levels for Na, K, and Cl to ppb levels for Sb and Au. This sequential analysis procedure shows great flexibility towards such a large range of concentrations, and the procedure can be adjusted, if necessary, at various stages to provide optimum sensitivity for specific elements.

Samples from eighteen sites in the NS&T Mussel Watch program have been analyzed with the sequential analysis scheme (XRF, PGAA, and NAA). A preliminary examination indicates some interesting results for particular sites, although factors such as species differences, etc. need to be considered before definite conclusions are drawn. Figure 4 shows a summary of concentration ranges among the sites for three elements: V (obtained by INAA), B (obtained by PGAA), and Pb (obtained by XRF-BFP). The sites in the figure are ordered geographically along the coastal United States, from Washington down the Pacific coast to the Gulf Coast, and then north along the Atlantic coast to Massachusetts. It should be noted that the Pb values for the samples are often at or very near the detection limit of the XRF-BFP technique. The values below the limit of detection are shown in Fig. 4 as the points with arrows. Even with many values at detection limits, analyses of such elements by nuclear methods

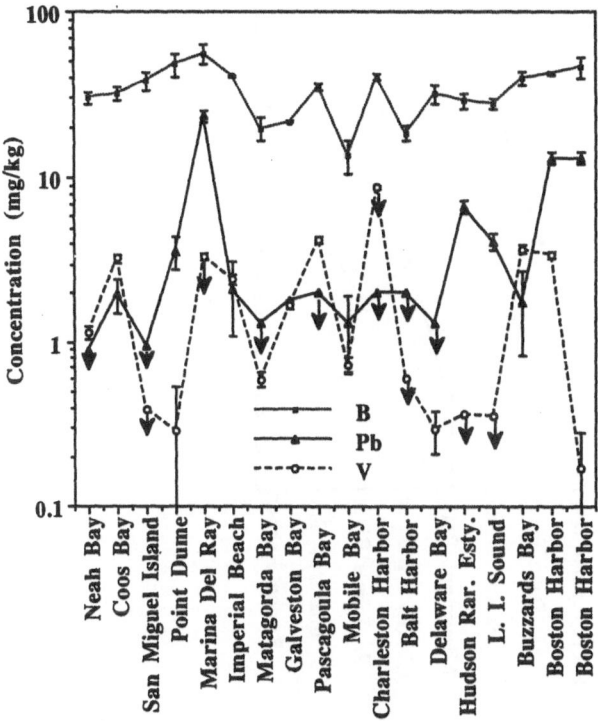

Fig. 4. Element concentrations (mg/kg, dry weight) for B, Pb, and V in 18 NOAA NS&T bivalve samples
Uncertainties represent individual uncertainties (1). Data below the limit of detection are represented as points with arrows

are valuable, because this information can give an idea of the upper levels present and show those sites which may have specific contamination.

Fish Livers

The Benthic Surveillance [17] portion of the NS&T program provides samples of bottom-dwelling fish and benthic sediment (discussed in the following section) to the NBSB. Both fish muscle and fish liver samples are collected, and selected samples of both of these tissue types were analyzed. The analysis of the fish liver samples contribute much more information, especially regarding possible pollutant elements, due to the liver's role in the detoxification of harmful compounds from organisms and the accumulation of these elements in the liver. Thus, the liver samples are preferred over the muscle tissue for analysis, even though it is necessary to dissect more fish to collect sufficient liver material (ideally 200–300 g).

The analytical scheme used to analyze the fish livers was similar to that of the human livers described previously. Concentrations for 33 elements were

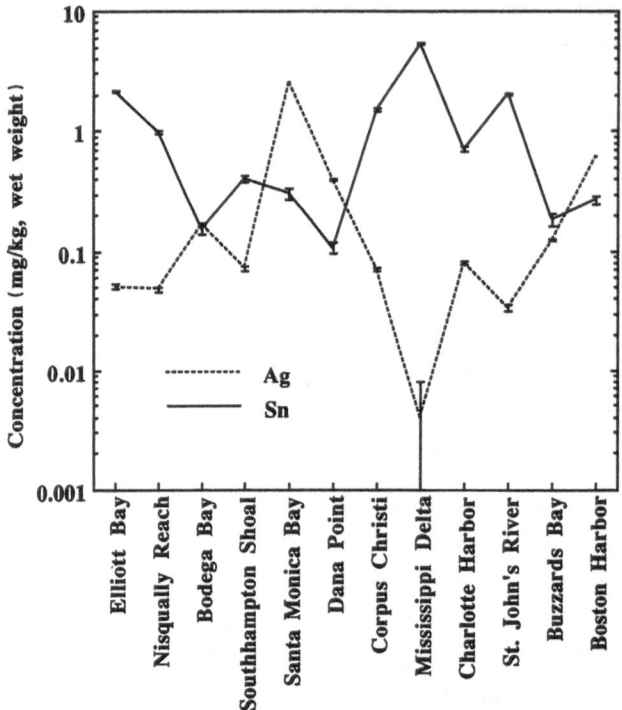

Fig. 5. Element concentrations (mg/kg, dry weight) for Ag and Sn in 12 NOAA NS&T fish liver samples
Uncertainties represent individual uncertainties (1)

Fig. 6a. Element concentrations (mg/kg, dry weight) for Ag and As in NOAA NS&T sediment samples
Uncertainties represent individual uncertainties (1). Data below the limit of detection are represented as points with arrows. **b** Enrichment factors for Ag and As in NOAA NS&T sediment samples, normalized to Al in soils

Site Identification:

A: Eliott Bay	Q: Mississippi River Delta
B: Nisqually Reach	R: Pasagoula Bay
C: Neah Bay	S: Charlotte Harbor
D: Coos Bay	T: St. John's River
E: Bodega Bay	U: Charleston Harbor
F: Southampton Shoal	V: Chesapeake Bay
G: Marina del Rey	W: Baltimore Harbor
H: Santa Monica Bay	X: Baltimore Harbor (SRM 1941)
I: Newport Bay	Y: Delaware Bay
J: Dana Point	Z: Long Island Sound
K: Imperial Beach	AA: Hudson Raritan Estuary
L: Corpus Christi	BB: Buzzards Bay
M: Gallinipper Point	CC: Goosebury Neck
N: Tres Palacois Bay	DD: Hingham Bay
P: Galveston Bay	EE: Dorchester Bay
	FF: Boston Harbor

obtained by INAA, and a final single element radiochemical separation was
used to obtain information on Sn values. Figure 5 gives a summary of data on
Ag and Sn for 12 fish liver samples. Due to their low levels, neither Ag and Sn
are analyzed frequently in biological samples, although they are significant
pollutant elements to monitor.

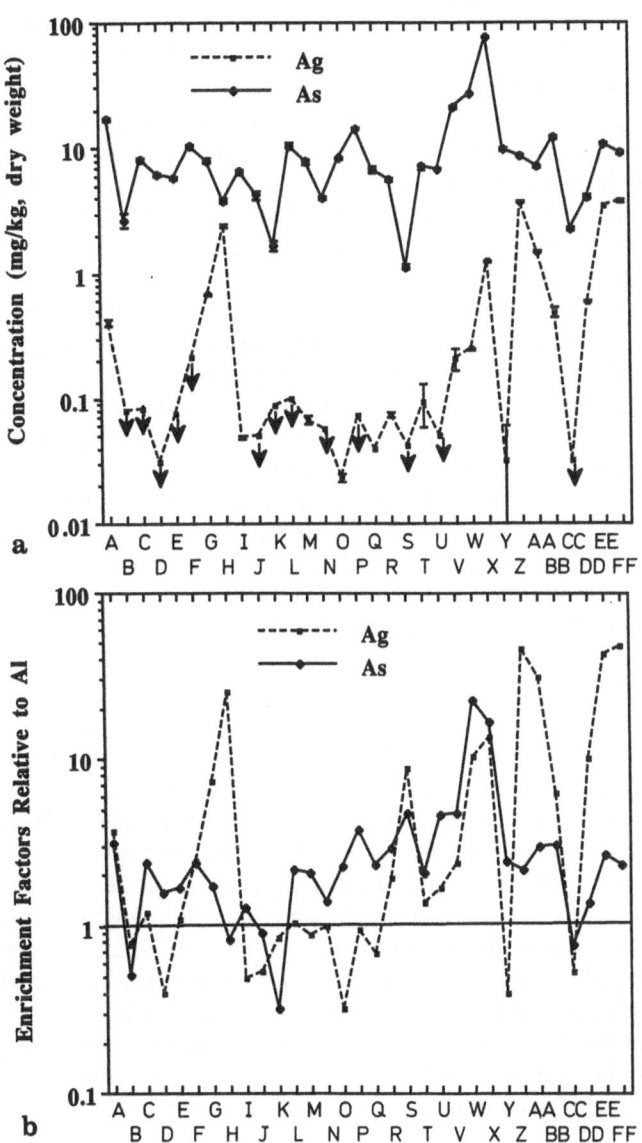

Sediments

Sediment samples were analyzed in conjunction with the bivalve and fish tissue samples. Since the sediments contain quite different concentration profiles compared to the biological samples, they require different analytical considerations compared to the analysis of biological samples. There were significant analytical problems for the trace and ultratrace pollutant elements of interest for the monitoring programs, caused by the elevated concentrations of many elements, such as the rare earth elements. Sediments also present difficulties for dissolution, and the analyses here are dependent totally on two instrumental techniques, INAA and PGAA. The details of the irradiation and counting conditions for these samples have been summarized previously [18, 19]. Loss-free counting techniques and other advanced instrumentation [18] figure prominently in the accurate analysis of these samples to obtain concentrations for up to 35 elements. A total of 18 Mussel Watch and 13 Benthic Surveillance sediment samples have been analyzed in the NBSB program. Efforts are underway to correlate concentrations of trace elements in biological and sediment samples from the same site.

A comparison of concentrations for two pollutant elements for 32 sediment samples is shown in Fig. 6a. The sites in the figure are again ordered geographically, and an additional site is included; one at Baltimore Harbor, where the material for SRM 1941 was collected [20]. The concentration profiles for Ag and As are given with their individual uncertainties (1σ). Data below the limit of detection are again represented as points with arrows.

Some interesting concentration profiles are observed in Fig. 6a. For instance, there is a very high Ag concentration at Marina del Rey, Long Island Sound, and Boston Harbor and high Ag and As concentrations at Baltimore Harbor. The profiles suggest interesting preliminary trends; however, for a realistic understanding of the possible pollutant trends, it is necessary to evaluate the results in a different manner. One technique for evaluation of such data is with the use of enrichment factors [21], where the elemental concentrations are normalized to a crustal element, such as Al, and compared to similarly normalized values for soils, sedimentary rock or comparable matrix. A comparison of enrichment factors (normalized to Al [22]) is illustrated in Fig. 6b. The indication of high concentrations at Marina del Rey, Long Island Sound, Baltimore Harbor and Boston Harbor are supported by the enrichment factors.

Data manipulations, such as the calculation of enrichment factors, are facilitated by employing nuclear techniques such as NAA for inorganic analyses since concentrations for a large number of elements can be obtained with the same aliquot. For cases where Al cannot be used for normalization, (if there is Al contamination, for instance), other crustal elements (e.g. Sc, Ce, Cs) are easily obtained by NAA and can be substituted. In using other analytical techniques, the flexibility of obtaining reliable concentrations for so many elements is not readily available, nor would such data manipulations be as straightforward.

Conclusion

The analysis of banked samples by nuclear methods provides an abundance of information, even in cases where limited sample sizes are available. A significant number of elements can be determined in a large range of banked specimens, especially when several nuclear methods are combined. The dynamic ranges, sensitivities, and multielement capabilities of three techniques, NAA, PGAA, and XRF, provide the high quality data needed in a banking program.

Acknowledgements: The authors wish to acknowledge the work of many individuals in the Center for Analytical Chemistry at the National Institute of Standards and Technology (NIST) who have been involved in the NBSB program, including E.S. Beary, K.A. Fitzpatrick, J.W. Gramlich, R.R. Greenberg, S.H. Harrison, W.R. Kelly, W.F. Koch, B.J. Koster, J.K. Langland, E.J. Maienthal, T.J. Murphy, L.J. Powell, T.M. Sullivan, and S.A. Wise. We also wish to acknowledge R. Sanders from Pacific Northwest Laboratories in Richland, WA, for the XRF analyses and M. Stoeppler and P. Ostapczuk from the Nuclear Research Center Jülich, Germany, for the voltammetry results. The work described in this chapter has been supported in part by the Office of Health Research, Office of Research and Development, U.S. Environmental Protection Agency; and the Oceans Assessment Division, National Oceanic and Atmospheric Administration.

References

1. Greenberg RR, Fleming RF, Zeisler R (1984) Intl Env Res 10: 129
2. Wise SA, Koster BJ, Parris RM, Schantz MM, Stone SF, Zeisler R (1989) Intern J Environ Anal Chem 37: 91
3. Failey MP, Anderson DL, Zoller WH, Gordon G, Lindstrom RM (1979) Anal Chem 51: 209
4. Nielson KK, Sanders RW, Evans JC (1982) Anal Chem 54: 1782
5. Sanders RW, Olsen KB, Weimer WC, Nielson KK (1983) Anal Chem 55: 1911
6. Zeisler R, Stone SF, Sanders RW (1988) Anal Chem 60: 2760
7. Zeisler R, Greenberg RR, Stone SF (1988) J Radioanal Nucl Chem 124: 47
8. Greenberg RR (1988) Anal Chem 58: 2511
9. Zeisler R (1983) In Zeisler R, Harrison SH, Wise SA (eds) The pilot national environmental specimen bank: analysis of human liver specimens, NBS Spec Publ 656. US Government Printing Office, Washington, DC 35 p
10. Zeisler R, Greenberg RR, Stone SF, Sullivan TM (1988) Fres Z Anal Chem 332: 612
11. Rossbach M, Zeisler R, Woittiez JRW (1990) Biol Trace Elem Res 26: 63
12. Zeisler R, Harrison SH, Wise SA (1984) Biol Trace Elem Res 6: 31
13. Iyengar GV, Kollmer WE, Bowen HJM (1978) The elemental composition of human tissues and body fluids. Verlag Chemie, Weinheim
14. Iyengar GV (1989) Elemental analysis of biological systems, volume 1: biomedical, environmental, compositional, and methodological aspects of trace elements. CRC Press, Inc., Boca Raton
15. Iyengar GV, Woittiez JRW (1988) Clin Chem 34: 106
16. Wise SA, Koster BJ, Parris RM, Schantz MM (in preparation)
17. Lauenstein GG, Wise SA, Zeisler R, Koster BJ, Schantz MM, Golembiewska SL (1987)

National Status and Trends Program for Marine Environmental Quality Specimen Bank Project: Field Manual, NOAA Technical Memorandum NOS OMA 37. US Department of Commerce National Oceanic and Atmospheric Administration, National Ocean Service
18. Stone SF, Koster BJ, Zeisler R (1990) Biol Trace Elem Res 27: 579
19. Stone SF, Becker DA, Koster BJ, Pella PA, Sleater G, Tillekeratne MPM, Zeisler R (1988) In: Wise SA, Zeisler R, Goldstein G (eds) Progress in environmental specimen banking. US Government Printing Office, Washington, DC, 62 p
20. Schantz MM, Benner BA, Chesler SN, Koster BJ, Hehn KE, Stone SF, Kelly WR, Zeisler R, Wise SA (1990) Fres J Anal Chem 338: 501
21. Zoller WH, Gladney ES, Duce RA (1974) Science 183: 198
22. Vinogradov AP (1959) The geochemistry of rare and dispersed chemical elements in soils, 2nd ed. Consultants Bureau Inc., New York, 163 p

6.2 Advanced Electrochemical Techniques for the Determination of Heavy Metals in Specimen Bank Materials

P. Ostapczuk, M. Froning

The possibility and limitation of electroanalytical methods such as stripping voltammetry and potentiometric stripping analysis for the determination of trace elements in samples from environmental specimen bank program are presented. High pressure digestion with nitric acid (HPA) is very useful as a sample digestion technique prior to voltammetric determination. For the PSA the digestion in PTFE vessels with nitric acid at 170 °C is sufficient. The accuracy of these methods has been demonstrated by the determination of Zn, Cd, Pb, Cu, Ni, and Co in different certified and Standard Reference materials and samples from the Environmental Specimen Bank.

Introduction

Toxic metals have become an area of particular concern and high priority in environmental research and protection. The most important feature distinguishing metals from other toxic pollutants is that metals are not biodegradable [1]. They are absorbed by humans from the environment via respiration and from the continental and marine food chains. The heavy metals such as lead, cadmium, mercury, arsenic, nickel, and selenium [2] have been reported to be occupational hazards, and some of them have been linked to cancer, heart diseases, or allergy. Various other trace metals e.g. copper, zink, cobalt, or tin have essential functions for living organisms below a critical concentration, but above which they begin to cause toxic effects. A thorough understanding of the effects of trace elements in the environment and in humans depends largely on the availability of sensitive and reliable analytical techniques.

Atomic absorption (AAS) [3], plasma emission spectroscopy (ICPAES) [4], electrochemical stripping analysis (ASV) [5], isotope dilution mass spectrometry (IDMS) [6], or neutron activation analysis (NAA) [7] are used for the determination of trace elements in samples from the Environmental Specimen Bank with varying degrees of success and convenience.

Recent advances in stripping analysis, including the adsorptive stripping voltammetry [8, 9], the development of various nonelectrolytic preconcentration schemes, potentiometric stripping technique, the introduction of modified and ultramicroelectrodes or of fast and versatile flow systems, have advanced the status of stripping analysis considerably [10]. The aim of this work is to demonstrate the possibilities and also the limitations of the electroanalytical

Specimen Banking
Rossbach/Schladot/Ostapczuk (Eds.)
© Springer-Verlag Berlin Heidelberg 1992

methods for the routine determination of some trace elements in samples from the German Environmental Specimen Bank (ESB).

Experimental

Equipment. For the voltammetric determination two different instruments were used: Model 384B Polarographic Analyzer with static mercury drop electrode (SMDE) Model 303A (EG&G) and Tracelab (ECO Chemie) with VA Stand 638 (Metrohm). Both instruments work in differential pulse and square wave mode. The potentiometric stripping measurements were done by TraceLab (Radiometer) in connection with SAM20 sample station (Radiometer).

Chemicals. All chemicals used (HNO_3, HCl, $HClO_4$, H_2O_2, CH_3COONa, KCl) were Suprapur (Merck) quality. The nitric acid was subboiled in quartz (Kürner, Rosenheim) before use. The water was prepared by Milli-Q system (Millipore).

Sample preparation. Figure 1 demonstrates the analytical scheme of ESB materials. For rain- and seawater, which are collected parallel to the Environmental Specimen Bank matrices, the sample preparation is simple. Lead and cadmium in rainwater samples can be determined directly by PSA without any sample

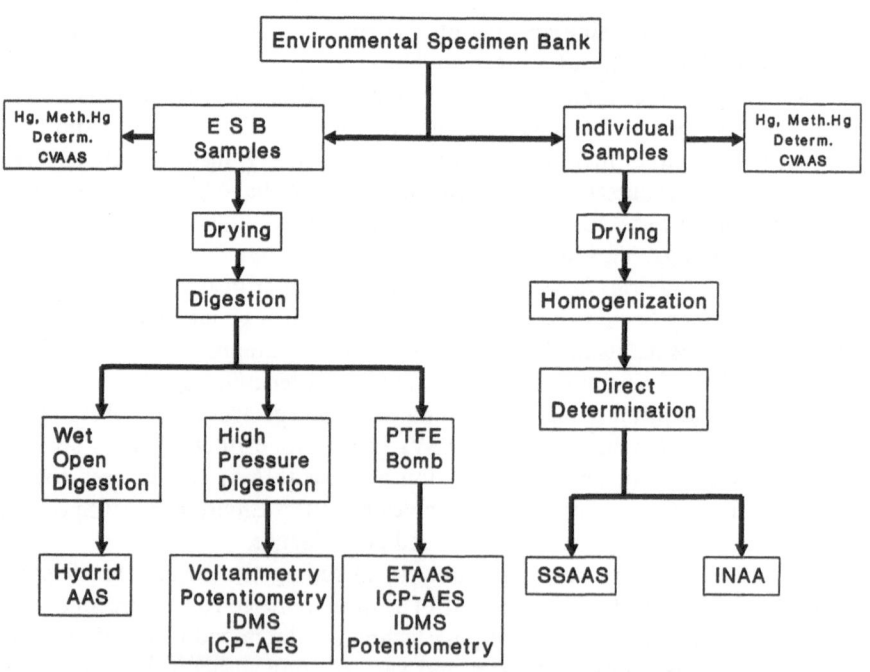

Fig. 1. Sample preparation prior to the analysis of individual and homogenous environmental specimen bank samples

preparation [11]. Prior to the voltammetric determination of these elements, the UV-irradiation [12] of seawater and rainwater samples was necessary to destroy any organic residues. The water filters and all other samples from the Environmental Specimen Bank were digested under high pressure with nitric acid in quartz vessels (HPA, Kürner, Rosenheim) at the maximum digestion temperature of 290 °C [13]. The total digestion time including cooling lasted 4 hours. In some cases (e.g. for Pt determination) the temperature was 320 °C and a mixture of nitric and hydrochloric acid was used. This acid mixture is also recommended if iron is to be determined in the samples. The frozen homogeneous and individual samples from the Environmental Specimen Bank were dried by lyophilization prior to digestion. Individual samples (e.g. earthworm, spruce shoot) were homogenized after drying by a ball grinder made from zirconium dioxide (Fritsch).

The analyte solution was prepared in two ways. For samples with a cadmium concentration higher than 100 μg/kg, only removal of dissolved NO_x by heating the digestion vessels at 100 °C (water bath) was necessary. In all other cases the evaporation of the digestion solution to dryness in quartz vessels was recommended. If the precipitate after evaporation has a brown or blue colour (iron or manganese compounds), the addition of 0.1 mL hydrochloric acid (30%) is necessary. The residue is dissolved in 0.1 mL of nitric acid and diluted to 10 mL with water. For very low concentrations (e.g. Cd and Pb in eggs) the residue is dissolved in 0.01 mL of nitric acid and 1 mL of water.

Voltammetric determination. 5 mL of supporting electrolyte (0.04 M perchloric acid) were degassed by passing argon through the solution for 4 min prior to, and over the solution during the experiment. Between successive runs a purging time of 30 s was used. The details of voltammetric determination are presented in [5, 14].

Potentiometric determination. 10 mL of rainwater was mixed with 0.1 mL hydrochloric acid, 0.1 mL of plating solution (Hg^{2+} 800 mg/L), and 0.5 mL 3.4 M CH_3COONa solution.

After 1–10 min (varying with the cadmium concentration) of preelectrolysis at -1.25 V vs. sat. Ag/AgCl, the stripping curve was measured by the computer system. The concentrations of cadmium and lead in the sample were evaluated by two standard additions. For digested samples an aliquot of acid digest was diluted to 10 mL in the electrochemical cell, using a solution containing 0.06 M hydrochloric acid and mercury(II) (Hg^{2+} 32 mg/L). The procedure for cadmium and lead determination was the same as described above for rainwater.

Results

Determination of Cadmium and Lead in Rainwater. The determination of heavy metals in atmospheric wet deposition is of particular interest in studies of the biogeochemical cycle of trace elements. It provides valuable information about heavy metal input from the atmosphere. Various methods such as ETAAS or TXRF are available for cadmium and lead determination in rainwater [15, 16].

In view of the large number of rainwater samples usually involved in a routine measurement program, and because of its high speed, simplicity, low cost, and sensitivity the differential – pulse or square wave stripping volta- metry after UV – digestion of the filtered samples is the method of choice [17].

Potentiometric stripping analysis (PSA) was also tested for the determina- tion of cadmium and lead in rainwater samples. Some of the filtered rainwater samples collected in our monitoring program were analyzed by DPASV and PSA. Table 1 demonstrates the comparison between the two techniques. The lead and cadmium concentrations found by both methods agree well in all cases. In the determination of Cd and Pb in rainwater PSA has the following advantages in comparison with voltammetry:

– the determination can be done without deaeration, because the oxygen is used in the chemical oxidation step,

Table 1. Comparison between voltammetric determination after UV-digestion and potentiometric direct determination in rainwater samples

Sample/date	Cadmium [ng/ml]			Lead [ng/ml]		
	UV-digestion		Direct	UV-digestion		Direct
	DPASV	PSA	PSA	DPASV	PSA	PSA
Deuselbach 26–30.11.90	0.194	0.206	0.198	2.29	1.48	1.55
Deuselbach 04–08.01.91	0.333	0.327	0.250	1.95	2.01	1.73
Bornhöved 15–22.01.91	0.153	0.162	0.149	6.35	6.02	6.47
Berchtesgaden 12–19.11.90	0.016	0.024	0.020	1.39	1.25	1.20
Schau ins Land 24–31.12.90	0.108	0.096	0.056	2.63	2.50	1.61
Welldorf 13–20.12.90	0.346	0.323	0.286	5.85	5.53	5.39
Welldorf 03–10.01.91	0.053	0.05	0.034	1.58	1.68	1.37
Essen 28.11–05.12.90	0.762	0.73	0.687	11.8	12.4	10.8
Essen 01–09.01.91	0.096	0.11	0.123	4.84	5.04	5.39
Binsfeldhammer 02–09.11.91	0.947	0.96	0.941	16.6	15.9	17.3
Binsfeldhammer 17–24.01.91	0.557	0.54	0.535	25.3	23.7	24.8

- cadmium and lead can be directly determined without any sample preparation of the rainwater sample, and
- the sensitivity of PSA is superior to the sensitivity of DPASV with SMDE.

Analytical experience has shown that when using PSA, the total analysis time, including sample preparation, can be reduced to about 15 min. The direct determination eliminates the contamination risk from vessels and chemicals by UV – digestion and the manpower needed for glassware cleaning.

In Table 2 the relation between cadmium and lead concentration in rainwater and filter samples from three different areas of Germany is presented. The highest cadmium content is found in rainwater. No significant variation in solubility of cadmium from filters from different collecting areas was observed. The solubility of lead from dust particles change strongly with the pH of rainwater. In all cases of decreased pH value ("acidic rain"), the lowest lead content on the filter was observed. Also an influence of the collecting area on the solubility of lead from particular matter was observed. By similar pH the solubility of lead in Bornhöved is better than the lead solubility in the industrial regions (e.g. Essen). It is presumed that the ionic composition of rainwater (high Cl^- concentration in Bornhöved and high SO_4^{-2} concentration in industrial areas) effects lead solubility similarly to pH. To elucidate this effect, more information about the ionic composition of rainwater is necessary. Figure 2 demonstrates the trend of lead concentrations in rainwater in three different areas of Germany, estimated from our experimental data. A significant decrease in the lead concentration in all collecting areas since 1981 was observed.

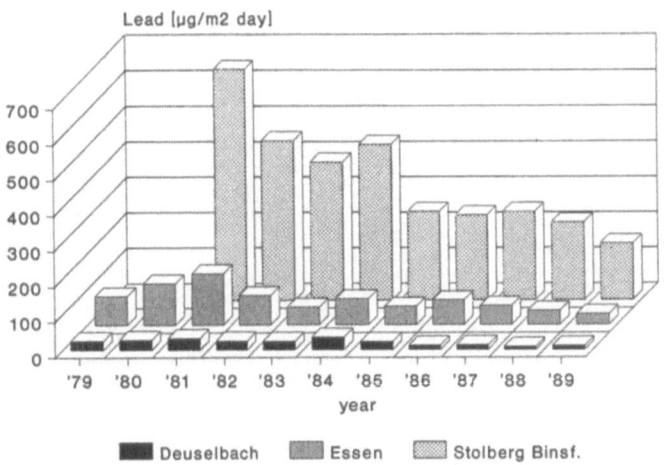

Fig. 2. Change of the total lead deposition in different region of Germany in last years: Deuselbach – rural region (unpolluted area); Essen – city region (pollution by car); Stolberg Binsf. – industry region (pollution by lead smelter)

Table 2. Influence of pH to solubility of cadmium in lead from dust

Week/year	Sample volume	pH	Cadmium			Lead		
	ml		Total ng/ml	Water %	Filter %	Total ng/ml	Water %	Filter %
Bornhöved:								
3/90	212	2.03	0.683	96.6	3.4	12.7	92.9	7.1
6/90	598	5.05	0.082	96.3	3.7	3.13	61.0	39.0
20/90	362	4.22	0.185	89.7	10.3	8.23	86.4	13.6
24/90	1716	6.36	0.026	92.3	7.7	1.02	68.5	31.5
Berchtesgaden:								
3/90	250	4.63	0.057	89.5	10.5	3.05	78.0	22.0
6/90	4985	5.14	0.058	92.4	8.6	1.04	38.5	61.5
20/90	950	4.76	0.134	97.8	2.2	3.34	75.1	24.9
24/90	3150	4.69	0.035	97.1	2.9	2.32	74.1	25.9
Essen:								
3/90	210	2.61	0.620	96.3	3.7	32.5	92.9	7.1
6/90	740	4.52	0.267	94.4	5.6	13.5	80.0	20.0
20/90	105	6.47	1.75	76.6	23.4	89.6	20.8	79.2
24/90	1290	4.34	0.205	99.0	1.0	9.29	70.8	29.2

Determination of Zn, Cd, Pb, Cu, Ni and Co in Seawater. Seawater is a complicated electrolyte system in which, besides several well-known macro-salt components, many heavy metals and less common metals are dissolved at the trace or even ultra trace level.

The dissolved metals interact with suspended particulate matter and sediments. Aquatic organisms, from the lowest trophic level (plankton) but also mussels and benthic organisms tend to accumulate toxic metals significantly [18].

Seawater and mussel samples from the North Sea were collected every two months for the monitoring program. The water and filter samples were analyzed immediately after collection. The mussel samples were stored at − 170 °C for future determination. The concentration range of zinc, cadmium, copper, nickel, and cobalt in filtered coastal seawater is similar to that observed sometimes in rainwater from unpolluted areas. Only the lead concentration in filtered seawater is significantly lower. Due to high pH value of seawater (7–8) the greatest part of lead is bound to particular matter. Figure 3 demonstrates the change of lead concentration with time in one of the collecting areas. Although the collected samples present only the momentary concentration, the characteristic change of the total content for the elements analyzed was observed over two years. The lowest values are found during the months with highest biological activity. In the winter months the highest total concentration of elements in seawater was observed. The trend of total lead concentrations in seawater during the last two years does not coincide with the trend observed in rainwater samples. Similar trends are observed in algae in this sampling area for longer time periods.

Determination of Trace Elements in Herring Gull Eggs. Birds are useful biological indicators for heavy metal pollutants in the environment due to their wide

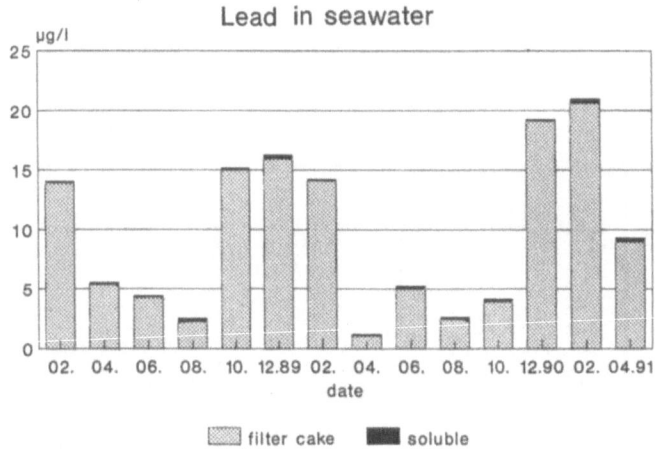

Fig. 3. Lead in seawater

Table 3. Zinc, copper, cadmium, lead, nickel, and cobalt concentrations in individual shells from sea gulls

No.	Zinc [mg/kg]		Copper [mg/kg]		Cadmium [µg/kg]	
	Mellum	Trischen	Mellum	Trischen	Mellum	Trischen
1.	1.83	1.05	3.18	1.65	2.80	3.66
2.	0.517	1.14	1.49	1.76	0.92	1.49
3.	0.236	0.710	2.02	1.99	1.60	2.47
4.	0.548	0.407	1.56	1.97	2.42	9.85
5.	0.963	0.421	2.37	1.69	1.20	5.61
mean:	0.818	0.746	2.12	1.81	1.79	4.62
s.d.;	0.622	0.343	0.69	0.16	0.80	3.30

No.	Lead [µg/kg]		Nickel [µg/kg]		Cobalt [µg/kg]	
	Mellum	Trischen	Mellum	Trischen	Mellum	Trischen
1.	68.9	24.1	59.9	93.0	9.16	5.13
2.	40.3	54.7	62.9	90.4	9.77	5.80
3.	45.7	43.1	183	84.1	7.44	5.83
4.	34.4	65.4	278	84.8	5.48	2.07
5.	42.1	67.5	470	173	9.49	12.7
mean:	46.3	51.0	211	105	8.27	6.30
s.d.:	13.3	17.9	171	38	1.80	3.90

distribution and higher trophic levels in the food chains [19]. Therefore, the anthropogenic influence on the heavy metals and organic contamination in seabirds along the North Sea should be monitored. Additionally, the North Sea coast supports large populations of seabirds and the pollution problems have been increasing [20, 21].

Egg samples of sea gulls (larus argentatus) were collected from two areas of the North Sea: Trischen and Mellum. To obtain more information about possible interaction between egg shell and the egg contents some individual shells were analyzed by voltammetry after HPA digestion with nitric acid. The high calcium concentration in the investigated solution had no influence on the voltammetric determination. The observed distribution of determined elements in the egg shells (Table 3) is characteristic for all environmental samples. Zinc followed by copper (both are essential elements) showed the highest concentration. Cadmium and cobalt concentrations are very low. Lead and nickel concentrations found in the shell are higher than concentrations found in the egg content. Contamination from these elements during sample collection and preparation is, however, possible.

Comparison between the concentration of zinc (Fig. 4a) copper (Fig. 4b), and cadmium (Fig. 5a) determined in individual eggs and in Environmental Specimen Bank samples as a composite of a large number of individual eggs, demonstrated that the concentration range observed in individual samples agrees very well with the concentration found in the homogenous composite sample.

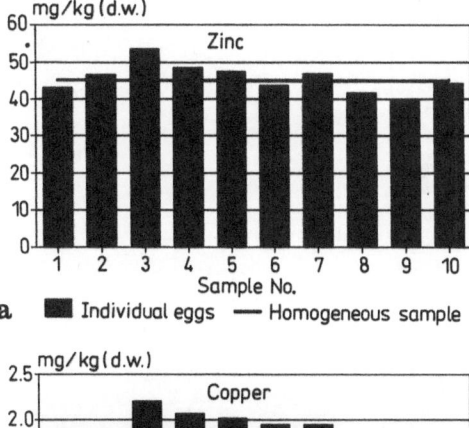

a   ■ Individual eggs — Homogeneous sample

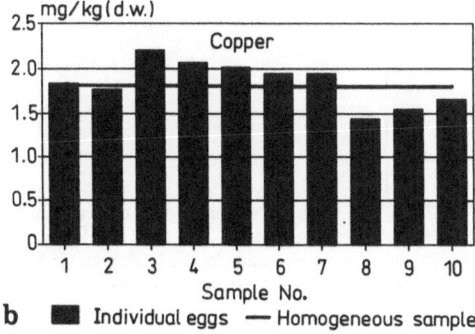

b   ■ Individual eggs — Homogeneous sample

Fig. 4. Zinc and copper concentration in individual samples and in environmental specimen bank samples

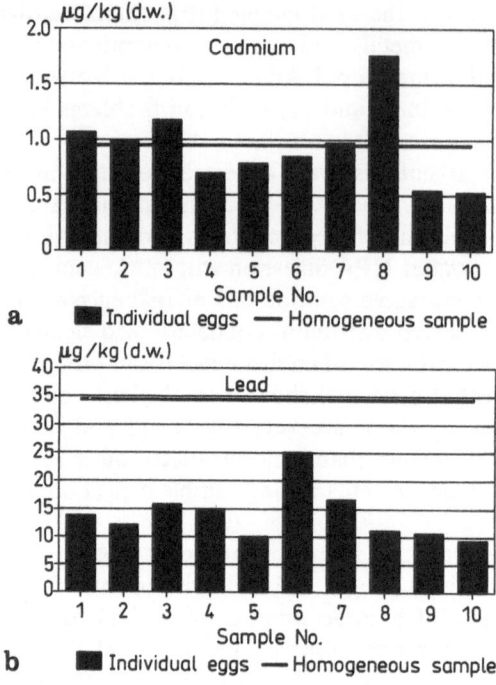

a   ■ Individual eggs — Homogeneous sample

b   ■ Individual eggs —Homogeneous sample

Fig. 5. Cadmium and lead concentration in individual samples and in Environmental Specimen Bank samples

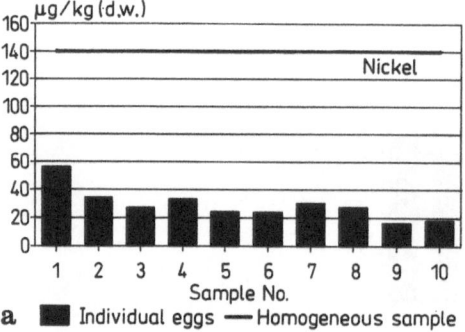

a   ■ Individual eggs —Homogeneous sample

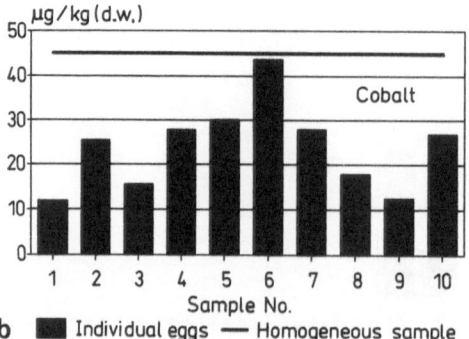

b   ■ Individual eggs — Homogeneous sample

Fig. 6. Nickel and cobalt concentration in individual samples and in Environmental Specimen Bank samples

Concentrations of lead (Fig. 5b), nickel and cobalt (Fig. 6a, b) determined in individual eggs are significantly lower than concentrations of these elements found in Environmental Specimen Bank samples. The results indicate that the homogeneous sample is contaminated by the above mentioned elements. More investigations to elucidate the possible source of contamination during the preparation procedure of homogeneous materials are necessary.

Determination of Elements in Spruce Shoots. Gravitational deposition, i.e. the deposition from rain, snow, dust etc. represent an important mechanism for the heavy metal input into the ecosystem. However, the deposition via interception, i.e. the impact of aerosols on the surface of vegetation, may contribute to a considerable extent. Especially over rough surfaces as are found in forest, input via interception may override that of gravitional deposition. The accumulation

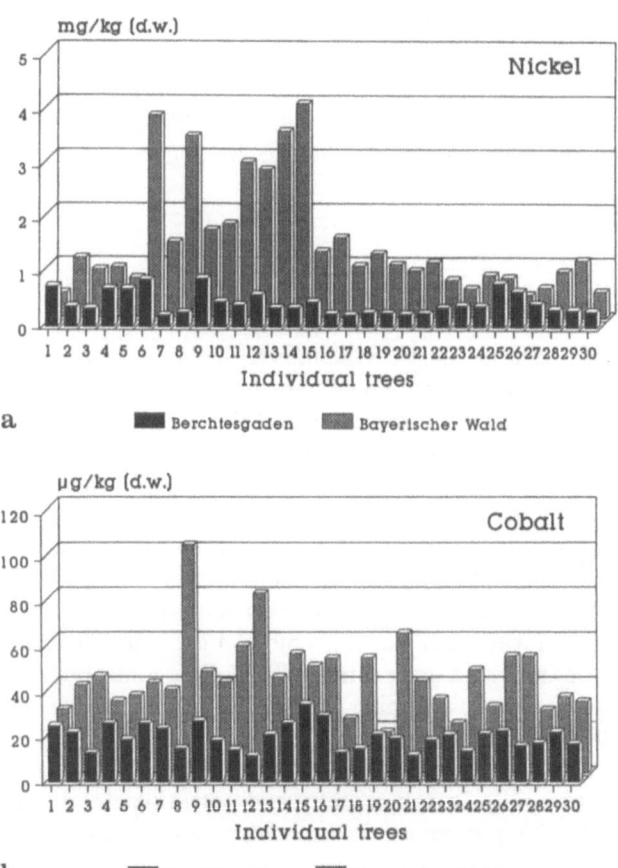

Fig. 7. Nickel and cobalt concentration in some individual samples from two regions: Berchtesgaden (unpolluted area) and Bayerischer Wald (polluted area)

of heavy metals is different in the various parts of a forest ecosystem. Low-level plants (especially lichens and mosses) can accumulate high amounts of metals. Higher plants, however, accumulate metals less actively because their cuticle-covered epidermis is not very permeable to metal gases [22].

Spruce shoots (picea abies) are one of the specimens collected within the German Environmental Specimen Banking program. In this matrix nickel and cobalt were determined by voltammetry. Figure 7 shows the concentration of these elements found in spruce shoots from 30 individual trees collected in Berchtesgaden and the Bayerischer Wald (the Bavarian Forest). The nickel concentration range is typical for all plant materials. In the samples from the Bayerischer Wald the nickel concentration in spruce shoots is considerably higher than the nickel concentration found in the spruce shoots from Berchtesgaden. More information about input of nickel to the atmosphere and the nickel content in the soil are needed to explain the sources of nickel accumulation. The natural cobalt concentration in plant materials is about 100 times lower than the nickel concentration observed and similar to concentrations found in eggs. The concentration range found in individual samples is for both elements similar.

The results of inorganic analysis in spruce needles from different regions of Germany demonstrated the usefulness of this matrix for the Environmental Specimen Banking program.

Conclusions

Electroanalytical methods such as inverse voltammetry or potentiometry are very sensitive methods for the determination of some trace elements in marine but also in terrestrial environmental samples. The delibrate selection of examples, which indicate the versatile applications and potential of electroanalytical methods to environmental chemistry of toxic trace metals, demonstrates the significance of these methods for this interdisciplinary field. The new developments in potentiometry demonstrates that this method can be also used for direct determination of zinc, cadmium, lead, thallium, copper and bismuth in water samples immediately after sample collection. Also inorganic mercury and arsenic can be determined in water samples. This method is predestined to be used in mobile laboratories for rapid environmental pollution control of some important elements. Potentiometric stripping analysis has significant advantages in comparison with voltammetry.

The new developments in computer software offers the possibility to automatize all the determination steps in voltammetry and potentiometry and the man power needed for the electrochemical determination can be reduced.

Acknowledgement: This study is part of the inorganic analysis carried out on the German Environmental Specimen Bank samples. The authors express their thanks to all their colleagues who have collected, prepared, and also analyzed the ESB samples.

References

1. Florence TM (1982) Talanta 29: 345
2. Ewers U (1984) In: Merian E et al. (eds) Metalle in der Umwelt, Verlag Chemie, Weinheim p 229
3. Stoeppler M, Backhaus F, Burow M, May K, Mohl C (1988) In: Wise SS, Zeisler R, Goldstein GM (eds.), NBS Special Publication 740, Progress in Environmental Specimen Banking, p 53, US Department of Commerce, National Bureau of Standards
4. Mohl C, Stoeppler M (1989) In: Welz B (ed) 5. Colloquium Atomspektrometrische Spuren-analytik, p 729, Bodenseewerk Perkin-Elmer
5. Ostapczuk P, Goedde M, Stoeppler M, Nürnberg HW (1984) Fresenius Z Anal Chem 317: 252
6. Hilpert K, Waidmann E (1986) Fresenius Z. Anal. Chem. 325: 141
7. Rossbach M (1990) In: Jül-Spez-576, ISSN 0343-7639
8. Wang J (1987) Analytical Proceedings 24: 325
9. van den Berg CMG (1989) Analyst 114: 1527
10. Wang J (1990) Fresenius Z Anal Chem 337: 508
11. Ostapczuk P (in preparation)
12. Mart L (1982) Talanta 29: 1035
13. Würfels M (1989) Marine Chemistry 28: 259
14. Ostapczuk P, Froning M, Stoeppler M, Nürnberg HW (1985) Fresenius Z Anal Chem 320: 645
15. Slanina J, Möls JJ, Baard JH, van der Sloot HA, van Raaphorst JG (1979) Intern J Environ Anal Chem 7: 161
16. Michaelis W (1986) Fresenius Z Anal Chem 324: 662
17. Nürnberg HW (1984) Analytical Chimica Acta 164: 1
18. Branica M (1990) In: Forschungszentrum Jülich (ed) Environmental research in aquatic systems
19. Honda K, Marcovecchio JE, Kan S, Tatsukava R, Ogi H (1990) Arch. Environ. Contam. Toxicol. 19: 704
20. Becker PH (1989) Helgoländer Meeresunter. 43: 395
21. Hahn E, Hahn K, Stoeppler M (1989) J. Orn. 130: 303
22. Krivan V, Schaldach G (1986) Fresenius Z. Anal. Chem. 324: 158

6.3 Determination of Trace Elements in Human Organs using ICP-AES

H.P. Bertram, C. Müller

Inductively coupled plasma atomic emission spectroscopy (ICP-AES) is integrated into the analytical scheme of inorganic sample characterization within the scope of the Environmental Specimen Bank (ESB) for Human Tissue Münster. The specific difficulties and the advantages of ICP-AES application in the performance of analyses of human matrices are pointed out. Detection limits for nonmetals, ET-AAS refractory elements, and reference ranges for strontium and barium in selected human specimens are given. Spectral interferences and possibilities to avoid them are discussed: e.g. background substraction for vanadium and beryllium, or determination of trace elements in tissue with phosphorus and calcium excess. For each human specimen stored in the ESB, all elements which may be detected by ICP-AES are listed together with the sample preparation. Furthermore, examples are given for application of the method for rapid qualitative diagnosis and consecutive quantitative treatment surveillance in acute inorganic intoxications with accidental trace elements.

Introduction

Inductively coupled plasma atomic emission spectroscopy (ICP-AES) has been well established for some years in the field of inorganic analysis. This method yields results with minimum sample preparation within short time. Other multielement methods (e.g. neutron activation analysis) require more complex equipment or a more complicated sample preparation and handling.

The development of highly sophisticated instruments extended the first applications of ICP-AES, in the field of metallurgical and geological analyses [1, 2], to the evaluation of environmental problems in biotic materials [3, 4, 5]. However, there was a lack of detailed information regarding the analyses of human organs. Within the scope of the Environmental Specimen Bank for Human Tissue Münster information on the stored sample is fundamental. Thus it was necessary to develop adequate methods for sample characterization. Inorganic sample data are collected by running an integrated analytical ICP-AES and atomic absorption spectroscopy scheme. Optimized ICP-AES conditions were obtained for each human specimen type with the goal being – minimum sample handling and as many reliable results as possible. Further-

The investigations were supported in part by the Federal Ministry of Research and Technology/the Federal Ministry of Environment, Nature protection and Nuclear safety/the Federal Environmental Agency of the Federal Republic of Germany.

Specimen Banking
Rossbach/Schladot/Ostapczuk (Eds.)
© Springer-Verlag Berlin Heidelberg 1992

more, ICP-AES may be used in rapid diagnosis and treatment survey of acute metal intoxications [6].

Equipment

The investigations were done with a Plasma II instrument (Perkin Elmer) equipped with the following options: 2 1-m Ebert monochromators: (A) 3600 ruling density (lines/mm)/spectral range 160–400 nm/resolution 0.009 nm/ linear dispersion 0.229 nm/mm, (B) 1800 ruling density/spectral range 160–800 nm/resolution 0.018 nm/linear dispersion 0.527 nm/mm; vacuum/ purge system; transfer optics area and torch compartment purged with nitrogen; RF power supply with a frequency of 27.12 MHz and a power output up to 1800 watts; plasma argon flow 5–25 mL/min; and system controlled by Perkin-Elmer 7500 Professional Computer with Motorola M68000 8-MHz microprocessor, a 40 megabyte Winchester hard disk and 2 floppy diskette drives each of 320 Kbytes capacity.

Detection Limits

Besides the extended linearity of the calibration curve, one of the important advantages of ICP-AES is the possibility of routine analyses of the following elements, which are inaccessible by, for example, electrothermal AAS:

a) Trace elements, which are refractory in ET-AAS because of the formation of stable carbides with the graphite tube. The scarcity of data about these elements in human tissue and body liquids reflect the analytical difficulties, although these elements possibly have a function in biological systems. With their presence in most bodily organs and in food, they fulfill two of the five conditions of essentiality for trace elements. Furthermore, an essential role for some of these elements in biotic systems has been proven (e.g. tungsten for goats) [7, 8, 9, 10]. Detection limits for some refractory metals are shown in Table 1. Data are given for aqueous solutions. Samples of digested tissue (nitric acid) imply higher limits (factor 2 to 5).

Table 1. ICP-AES detection limits for some refractory metals

Element	Detection limit (μg/L)
Lanthanum	10
Titanium	1.0
Tungsten	20
Yttrium	2.0
Zirconium	5.0

b) Some nonmetals. In contrast to atomic absorption (AAS) the wavelength range 200–170 nm is well accessible to the ICP-AES system, due to nitrogen purging of the optical system. Table 2 lists the detection limits of the bulk elements sulphur and phosphorus and the trace elements boron, silicium and iodine.

c) Group IIa elements. The high sensitivity of the ICP-AES system for these elements is comparable to that of the electrothermal AAS (ET-AAS) (Table 3). Hence, it was possible to collect sufficient data (n 1300–1450) to give reference ranges for the trace elements strontium and barium in some human specimens (Table 4).

Due to convenient detection limits for other essential trace elements (Table 5), zinc, copper, and manganese, for example, may be determined in most human specimens. Analysis of physiological concentrations of accidental elements such as cadmium, lead, or arsenic with most systems, is restricted to tissue

Table 2. ICP-AES and ET-AAS detection limits for some nonmetal elements

| Element | Detection limit ($\mu g/L$) | |
	ICP-AES	ET-AAS
Phosphorus	50	200
Sulphur	50	–
Boron	5	20
Silicium	10	0.5
Iodine	100	–

Table 3. ICP-AES and ET-AAS detection limits for group IIa elements

| Element | Detection limit ($\mu g/L$) | |
	ICP-AES	ET-AAS
Beryllium	0.5	0.1
Magnesium	0.1	0.05
Calcium	0.15	0.05
Strontium	0.5	2.0
Barium	0.2	0.2

Table 4. Reference ranges for strontium and barium in human specimens gained by ICP-AES

| Specimen | Reference ranges | | | n |
	Strontium	Barium		
Blood plasma	(15) 20–90 (120)		$\mu g/L$	1450
Urine	(20) 30–250 (350)	(0.5) 1.5–5.0 (10.0)	$\mu g/L$	1450
Scalp hair	(0.5) 1.0–5.0 (12)	0.20–1.0	$\mu g/g$	1300

Table 5. ICP-AES and ET-AAS detection limits for some essential trace elements

Element	Detection limit (μg/L) ICP-AES	ET-AAS
Chromium	5.0	0.01
Iron	1.0	0.05
Copper	2.0	0.02
Manganese	0.2	0.005
Zinc	0.5	0.001

types with enriched contents (e.g. Cd in the kidney). In contrast ICP-AES detects the elevated levels in acute intoxications of most metals in tissue and body liquids (see below).

Spectral Interferences

One of the most striking difficulties of plasma emission spectroscopy is the problem of spectral interferences. The high torch temperature of up to 10 000 K results in an emission spectrum of high complexity. Argon, solvent and sample emissions must be taken into consideration to avoid overlapping of single lines. Thus, it is necessary to check each element line of each specimen type for interferences and choose another wavelength in the case of unavoidable problems. Table 6 lists the optimized analytical conditions for some human specimens.

Background substraction: Application of some sensitive element lines in analyzing biological matrices involves, in some cases, the necessity of a background substraction technique, otherwise this line is useless. This is important for analyses performed in wavelengths regions where OH and NO bands are present. After evaporation of the solvent or blank solution the software substracts the peaks within the chosen spectral window from the emissions of the following sample solutions. Figure 1 demonstrates this method: the spectral 0.080-nm-window with the most sensitive vanadium line at 310.230 nm shows two background emission lines at 310.217 and 310.239 nm (a), which are not differentiated totally from the V emission, even by monochromators with high resolution. After background substraction (b) a detection limit of 10–20 μg V/l may be reached.

Similar conditions are present for the sensitive beryllium line at 313.042 nm (Fig. 2a). A detection limit of ≈ 2 μg Be/l is possible without background correction. A suitable alternative is the application of the Be line at 234.861 nm without interfering background emission and a detection limit of ≈ 0.5 μg/l (Fig. 2b).

Element excess: High content of bulk elements in human specimens may result in spectral interferences not only with the prominent lines of these

Table 6. ICP-AES analytical conditions for determination of trace elements in selected human specimens

Element	Wavelength (nm)	Blood plasma	Urine	Human milk	Hair	Soft tissue	Bone tissue	Remarks
Copper	324.754	x		x	x	x	x	
Iron	238.204	x	x	x	x	x	x	
Zinc	213.856	x	x	x	x	x	x	
Strontium	407.771	x	x	x	x	x	x	bone 346.446 nm
Barium	455.403		x		x	x	x	
Manganese	257.610		x		x	x	x	
Boron	249.773				x	x	x	liver/kidney
Cadmium	214.438				x	x		

Sample preparation: blood plasma: direct dilution 1 + 4/urine: native, without dilution/human milk: direct dilution 1 + 1 with 0.2% homogenizer/hair, soft and bone tissue: nitric acid digestion, dilution 1 + 2 to 1 + 4

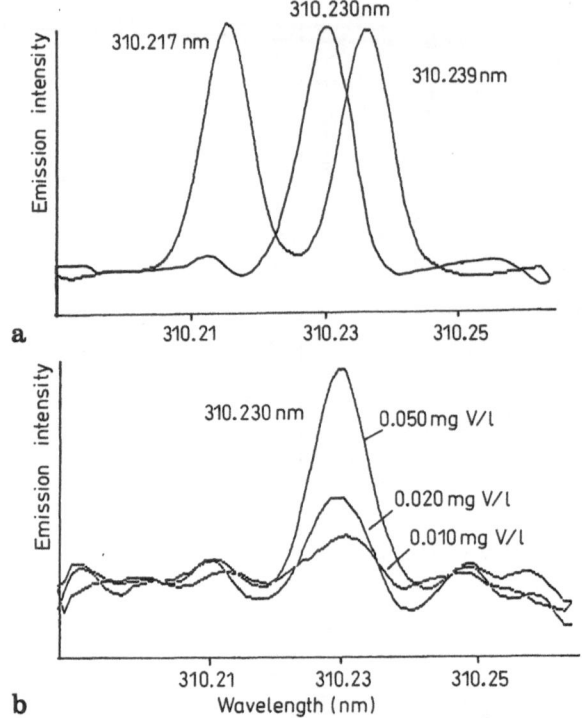

Fig. 1a, b. Vanadium emission line at 310.230 nm: a with two interferring background emissions (310.217 and 310.239 nm); b after background substraction

components, but also with secondary lines of normally lower importance. Trace element analysis of bone tissue is strongly influenced by the emissions of the bulk elements phosphorus and calcium. Selection of useful element lines is of particular importance. The strontium line at 215.284 nm and a secondary P line at 215.294 nm cannot be separated. In contrast the strontium wavelength at 430.545 nm may be useful: Fig. 3 shows the 1.5-nm-window around this line (1 mg Sr/l) at an excess of phosphorus (1000 mg P/l) and calcium (1000 mg Ca/l).

The possibility of trace analysis of iodine in human specimens with ICP-AES is limited by the normally high content of phosphorus in these samples. The most sensitive emission at 178.276 nm interferes strongly with the P emission at 178.283 nm. There are two ways to solve this problem: (1) usage of the mentioned line in second order (I 356.552 nm/P 356.566 nm), (2) usage of the second prominent iodine line at 183.038 nm. Unfortunately both possibilities are connected with diminished sensitivity. Without further enrichment iodine may be determined in thyroid gland tissue at 183.038 nm.

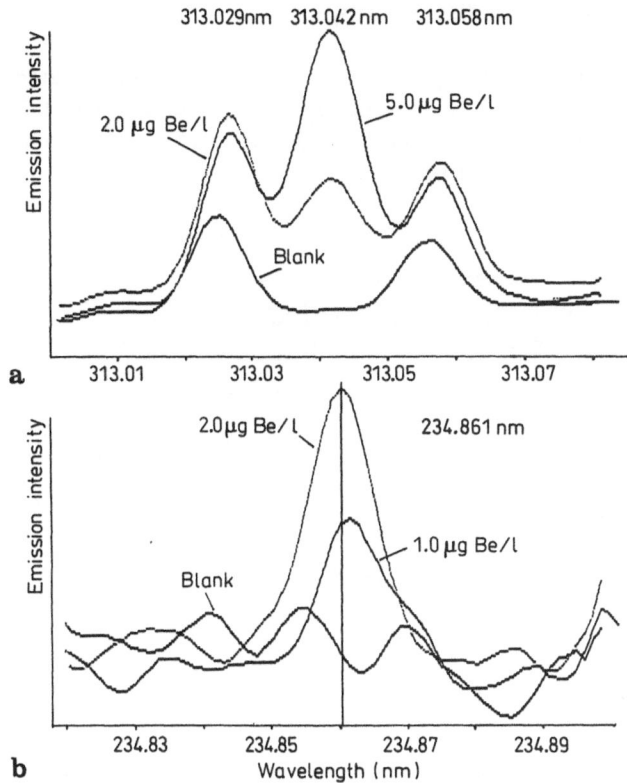

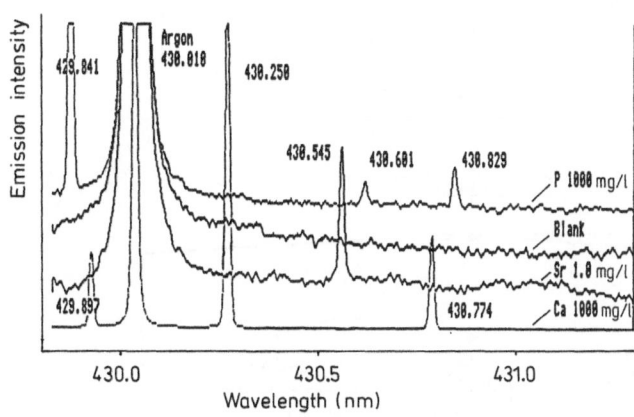

Fig. 2a, b. Most sensitive beryllium emission lines: a at 313.042 nm and b at 234.861 nm

Fig. 3. 1.5-nm-window at the 430.545 nm strontium emission line (1 mg Sr/L) with an excess of phosphorus and calcium (each 1000 mg/L)

Characterization of Specimen Bank Samples

ICP-AES is integrated into the analytical schemes for inorganic sample characterization for the Environmental Specimen Bank for Human Tissue Münster. In addition the different variations of AAS are included (ET-AAS, cold vapour (CV-) AAS, hydride generation (HY-) AAS).

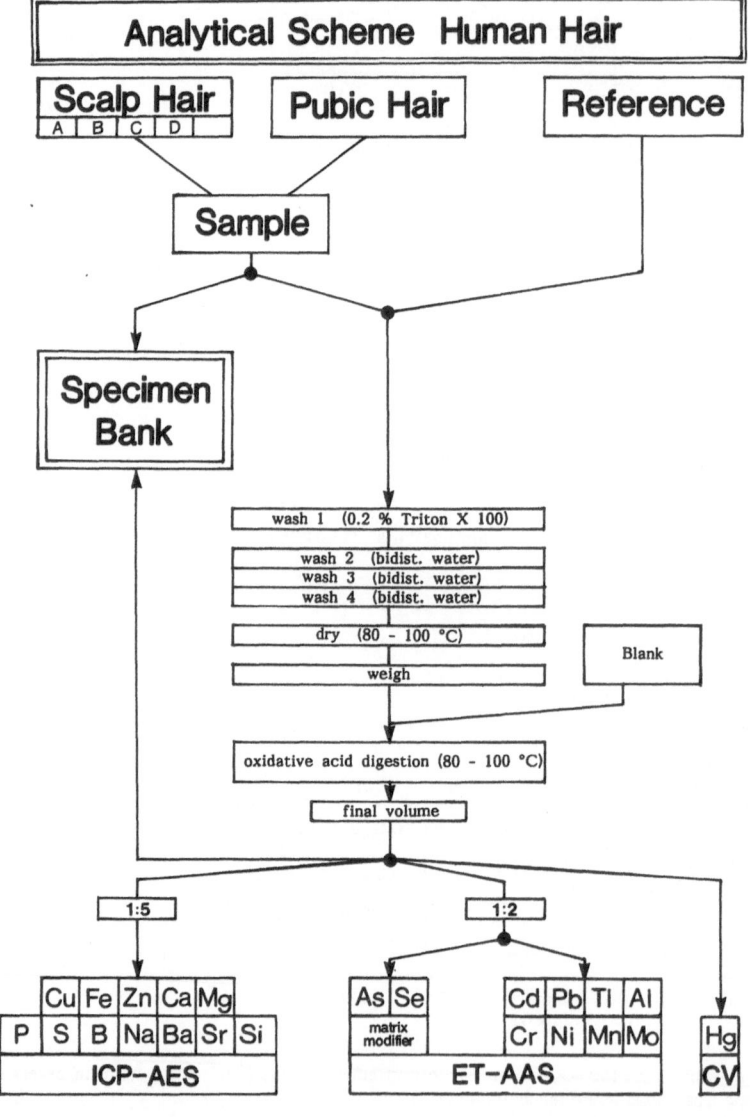

Fig. 4. Sample characterization scheme for human hair in the Environmental Specimen Bank for Human Tissue Münster

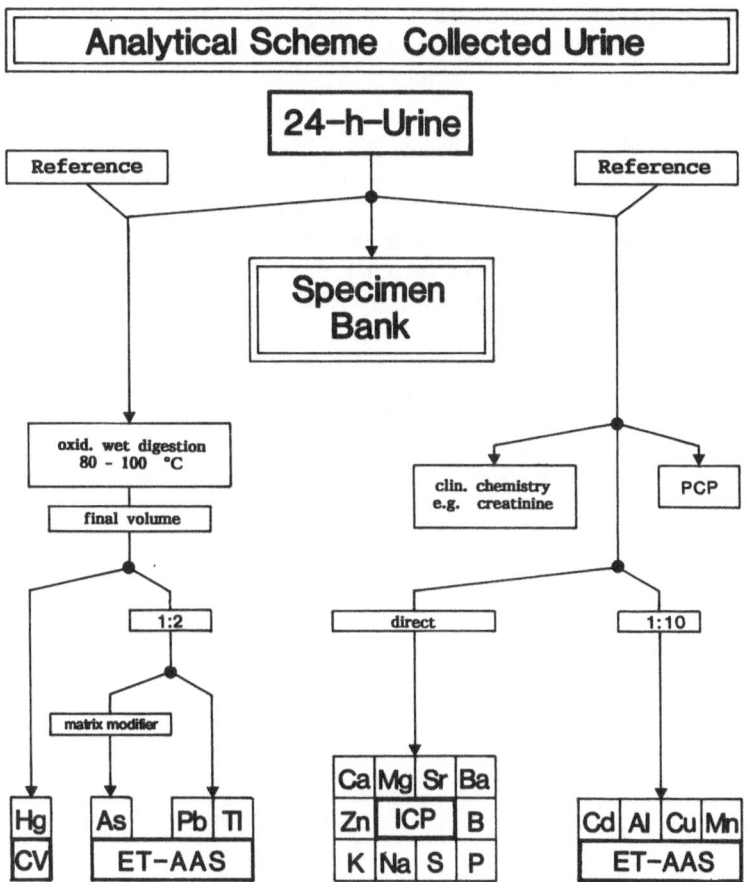

Fig. 5. Sample characterization scheme for human urine in the Environmental Specimen Bank for Human Tissue Münster

In human hair the trace elements copper, iron, zinc, strontium, barium, and boron may be determined routinely by ICP-AES (Fig. 4). If the sample size is sufficient (>200 mg for Mn/ >500 mg for Cd), analysis of manganese and cadmium is also possible [11]. In native undiluted urine the contents of zinc, iron, strontium, barium, and in most samples manganese, may be gained without further sample handling (Fig. 5). ICP analytical conditions for selected human specimens (blood plasma, urine, human milk, hair, soft and bone tissue) are given in Table 6. Analytical schemes for the analysis of whole blood, blood plasma and human milk are given in Chap. 2 (F.H. Kemper).

Application of ICP-AES in Acute Intoxications

ICP-AES is the method of choice for rapid qualitative diagnosis and consecutive treatment surveillance of acute inorganic intoxications. Toxic concentrations of trace elements in human body fluids are significantly higher than physiological contents and in nearly all cases are detectable by ICP-AES. Wavelengths and detection limits of the urine screening sequence in suspected metal overexposure are listed in Table 7.

Table 8 demonstrates some examples of identification and quantitation of inorganic poisonings. All data are results of analyses of native urine without dilution: (a) Cobalt: 14 years old boy. Ingestion of an unknown substance from an experimental chemical set. (b) Mercury: female, 32 years, 3 years after intravenous application of 2 g metallic mercury. (c) Cadmium: male, 37 years, chronic inhalative exposure at work place. (d) Antimony: child, 1 4/12 years,

Table 7. ICP-AES wavelengths and detection limits for a urine screening sequence in metal intoxications

Element	Wavelength (nm)	Detection limit (μg/L)
Aluminium	396.152	25
Antimony	206.833	50
Arsenic	193.696	30
Barium	455.403	0.2
Bismuth	223.061	50
Cadmium	214.438	2.0
Chromium	205.552	5.0
Cobalt	238.892	10
Copper	324.754	2.0
Lead	220.353	50
Lithium	670.784	5.0
Manganese	257.610	0.2
Mercury	194.227	50
Molybdenum	202.030	5.0
Nickel	231.604	10
Thallium	276.787	20
Tin	189.980	30
Vanadium	292.402	5.0

Table 8. Identification and quantitation of acute poisonings with inorganic substances by ICP-AES (native urine)

Element		Wavelength [nm]	Emission intensity	Concentration [μg/L]	Normal reference range [μg/L]
a	Cobalt	238.892	64526.97	26400	< 0.1–3.0
b	Mercury	194.227	394.50	3065	< 0.1–2.0
c	Cadmium	214.438	179.10	58.0	< 0.5–1.0
d	Antimony	206.833	376.80	12373	< 1.0
e	Barium	455.403	49237.10	17278	1.5–5.0
f	Thallium	276.787	583.60	5930	< 0.1–2.0
		190.864	223.70	6070	

accidental ingestion of an unknown amount of the formicide antimony potassium tartrate (tartar emetic). (e) Barium: male, 15 years, suicidal ingestion of 30 g barium chloride. (f) Thallium: female, 24 years, suicidal ingestion of an unknown amount of a rodenticide containing 2% thallium sulphate (results of different Tl wavelengths).

Conclusion

ICP-AES is a convenient and adequate multielement detection method and may be integrated into an analytical scheme in addition to e.g. ET-AAS for inorganic sample characterization of human samples within the scope of the Environmental Specimen Bank for Human Tissue Münster. It completes the analytical data, yielding results for some nonmetals (sulphur, phosphorus, boron, and iodine) and for ET-AAS refractory elements like titanium, tungsten or zirconium. Desirable minimum sample handling is given especially for body liquids, which may be analyzed directly with or without dilution. Reliable results with sufficient precision and accuracy are obtained only, if the emission lines for each element in each specimen type are thoroughly selected. The most important difficulties are spectral interferences with OH, NO, Ar, and other background emissions. In human specimens secondary lines of the bulk elements phosphorus and calcium may also result in erroneous data. Instrument development with a mass spectrometric detector system (ICP-MS) may increase the importance of plasma emission spectroscopy in the future [12].

References

1. Sulcek Z, Rubeska I, Sixta V, Paukert T (1989) At Spectr 10: 4
2. Batistoni DA, Faria de Funes SS, Smichowski PN (1990) Inductively Coupled Plasma-Atomic Emission Spectrometric Determination of Magnesium and Manganese in Aluminum and Aluminum Alloys. At Spectr 11: 85
3. McQuaker NR, Kluckner PD, Chang GN (1979) Calibration of an Inductively Coupled Plasma-Atomic Emission Spectrometer for the Analysis of Environmental Materials. Anal Chem 51: 888
4. Schramel P, Klose BJ, Hasse S (1982) Die Leistungsfähigkeit der ICP-Emissionsspektroskopie zur Bestimmung von Spurenelementen in biologisch-medizinischen und in Umweltproben. Fresenius Z Anal Chem 310: 209
5. Holm J, Müller S (1987) Multielementanalyse mit Hilfe der ICP in Lebensmitteln tierischer Herkunft. Fleischwirtsch 67: 466
6. Bertram HP (1990) Aufnahme eines Spurenelement-Rasters in Humanproben – Ökotoxikologische und medizinisch-klinische Bedeutung. Habilitation thesis, University of Münster
7. Schwarz K (1977) Essentiality versus toxicity of metals. In: Brown SS (ed) Clinical chemistry and chemical toxicology of metals. Elsevier/North-Holland, Amsterdam, p 3
8. Geldmacher-v.Mallinckrodt M (1984) Bedeutung der Schwermetalle in der Humantoxikologie. Fresenius Z Anal Chem 317: 427
9. Nielsen FH (1982) Possible future implications of nickel, arsenic, silicon, vanadium, and other ultratrace elements in human nutrition. In: Clinical, Biochemical, and Nutritional Aspects of Trace Elements. Alan R. Liss, New York, p 379

10. Kieffer F (1991) Metals as essential trace elements for plants, animals, and humans. In: Merian E (ed) Metals and their compounds in the environment. VCH, Weinheim, p 481
11. Bertram HP, Müller C (1987) Umweltprobenbank für Humanorganproben Münster – Haaranalysen. In: Krause C, Chutsch M (eds) Haaranalyse in Medizin und Umwelt. Gustav Fischer, Stuttgart, p 71
12. Lyon TDB, Fell GS, Hutton RC, Eaton AN (1988) Evaluation of Inductively Coupled Argon Plasma Mass Spectrometry (ICP-MS) for Simultaneous Multielement Trace Analysis in Clinical Chemistry. J Anal Atom Spectrom 3: 265

6.4 Future Prospects of Graphite Furnace Atomic Absorption Spectrometry

R.F.M. Herber

Since the introduction of graphite furnace atomic absorption spectrometry in 1961, the development of the technique pointed into two directions. One direction is mainly carried out by the users, i.e. (matrix) modification of the test substance. In fact, the test substance is modified in such a form, that it can be determined by the apparatus available. Although from the scientific viewpoint the generalization should be as wide as possible, in practice this is very rare. In the majority of the cases a modification procedure is described exclusively applicable for one element in one matrix. Thus this direction hardly adds new information about the technique as such.

The other development is the improvement of the technique by modification of the apparatus. In principle, this direction is more promising as improvements will be matrix and element independent. Instrumental improvements can be carried out both by users, if their laboratories are equipped for the purpose, and by manufacturers of apparatus. It is argued that the most important improvements are or will be the type of furnace and furnace material, the control of the temperature program, the signal analysis system, and the background correction system. With these improvements, more reliable determinations with a better precision and a better sensitivity can be obtained.

Introduction

The flame atomic absorption spectrometry (F-AAS) technique, as introduced by Walsh [1] and Alkemade [2] many years ago is well established and AAS apparatus are found most frequently in the laboratory, together with spectro-photometers and gas chromatographs.

Since the introduction of the graphite furnace (GF) in AAS by L'vov [3] detection limits went down dramatically, and especially determination of toxic metals in biological and environmental specimens was more feasible than before. However, it also became clear that problems connected with the matrix composition of the test specimen could be severe. Hence, different procedures were followed to overcome the matrix problem, i.e.

1. Complete destruction of the matrix by acid digestion followed by sub-sequential complexation and extraction of the analyte
2. Modification of the matrix within the furnace through the use of a catalyst such as palladium or chemical compounds, ensuring a less volatile analyte component, thus separating the analyte from the nonanalyte
3. If the problem is only spectrometric, correction is performed by a background correction system of the nonanalyte peak appearing during atomization

Specimen Banking
Rossbach/Schladot/Ostapczuk (Eds.)
© Springer-Verlag Berlin Heidelberg 1992

4. The introduction of instrumental improvements, ensuring separation of analyte and nonanalyte peaks under exactly repeatable conditions

It must be stated, that the introduction of the personal computer (PC) in GF-AAS as in other techniques was also effective. Commercial manufacturers are more directed towards using a PC for easier handling of the instrument and data handling than for improvement of the analytical characteristics of the technique, i.e. sensitivity, resolution of the analyte from the nonanalyte peak, detection limit, accuracy, and precision. Moreover, the introduction of a PC with color-screen seems to be a more important sales argument than a lower detection limit. Hence, the different analytical parameters will be critically reviewed in this report, and indication will be given, on the results of experiments in our laboratory, how improvement of these parameters may be obtained by instrumental improvements, including the use of a PC.

Drawbacks of Chemical Modification of the Matrix; Advantages of Instrumental Improvements

From the scientific point of view, every method should be as general as possible. Therefore, a chemical modification of the matrix for only one element of one matrix is unacceptable. Thus, the matrix modification by a destruction technique or in the GF should be developed in such a way, that a row of elements can be determined in different matrices. This, however, is rarely the case. Numerous modifications are needed, especially, in matrix modification in the furnace and in extraction techniques to ensure an accurate determination.

Another drawback for both matrix modification in the furnace and digestion techniques is the possibility of contaminating some elements. The more test specimen steps are needed, the higher is the risk of contamination. Moreover, the precision is also a negative function of the number of test specimen steps. It is therefore stated, that instrumental modifications that avoid the above mentioned chemical modifications of the matrix are to be preferred.

General Outlines of AAS Instruments

AAS instruments basically consists of five major systems, i.e. the lamp, the furnace, the detector, the signal analysis, and the background correction. Of these five components research has been mostly directed into the furnace and hence this system will be reviewed here. There are many commercial modifications of the background correction system and this system will also be reviewed. Finally, the signal analysis component will be reviewed, because by the improving speed and capacity of PC's, this part of the system should yield the most promising results.

The general approach should be sample directed, i.e. a sample in its physical and chemical entity has to be accepted without chemical modifications or physical phase modifications. This statement has important implications on the instrumentation as will be shown later on.

This review will concentrate on the critical points, rather than give a comprehensive report.

The Furnace

As previously stated, the general approach should be test specimen-directed. In the case of the furnace, this means that a separation can be made between liquid and solid samples and, consequently, in furnaces for liquids and furnaces for solids. The reason for this difference is that liquids contain relatively small amounts of solid material, when compared with the solid samples. In the case of solids, an enormous smoke cloud may appear. If the furnace is too small, and the gas transport too low, the relatively dense cloud will stay in the furnace, thus giving problems for the optical transmission. Moreover, the large amount of solid material needs a greater expansion volume during atomization. Figure 1 shows a furnace specially developed for solid sampling analysis. A disadvantage of such a furnace is that a large power unit is required, with sometimes a three phase power line. An advantage of the furnace is that no injection hole is needed, as the solid sample transport takes place with the aid of a pincette or a solid sampler transport system [4]. It has been reported, that with the injection hole in the furnace, up to 30% losses of the atom cloud may occur. Because of the

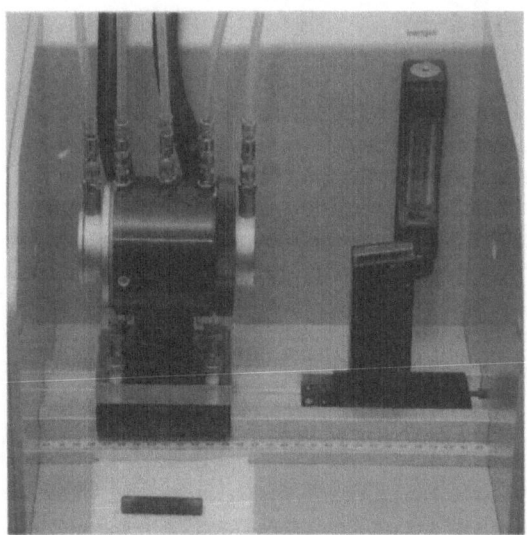

Fig. 1. Large size furnace for solid test substances, with longitudinal heating system manufactured by Grün Analysengeräte, G.m.b.H., Wetzlar, Germany; Type SM20
Dimensions: length = 53 nm, O.D. = 10.5 mm, I.D. 8.5 mm, and weight = 2670 mg

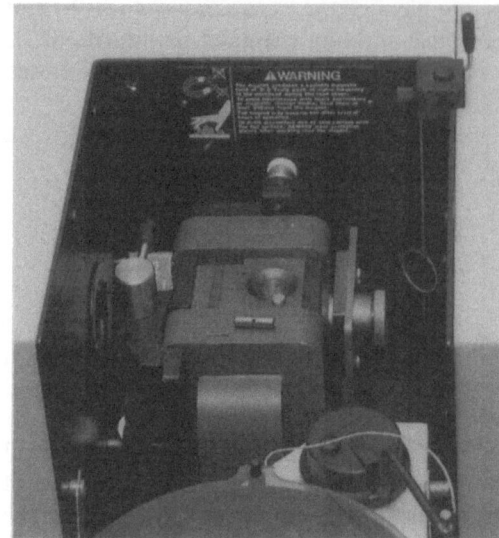

Fig. 2. Medium size furnace for liquid test substances, with longitudinal heating system manufactured by Varian Techtron Ltd, Pty, Melbourne, Australia; Type GTA96
Dimensions: length = 25 mm, O.D. = 7 mm, I.D. = 5.2 mm, and weight = 830 mg

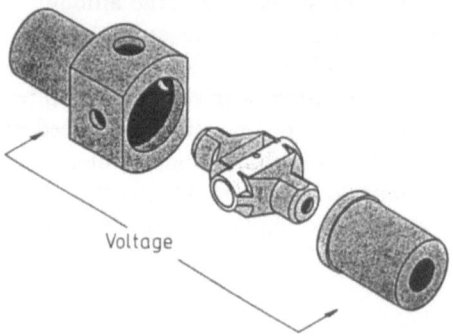

Voltage

Fig. 3. Medium size furnace with transverse heating system, manufactured by Perkin Elmer Corp, Norwalk, U.S.A.; Type 4100 ZL

smoke development with the furnace, a very efficient background correction system is needed, i.e. the Zeeman effect system.

Furnaces for liquid samples are smaller than for solids, as the amount of solid material after evaporation is less. A typical example of such a furnace is shown in Fig. 2. A disadvantage of this type of furnace is that they are longitudinally heated, and this heating system in combination with the length is responsible for cooler ends of the furnace. This again, leads to condensation problems and a low separation between analyte and nonanalyte peaks. To overcome this, a transverse heated furnace was developed [5]. This system is now commercially available and should be advantageous over the older furnaces, see Fig. 3. With the above mentioned "normal" furnace, most elements can be successfully determined in liquids.

However, for some elements another furnace which is smaller may be advantageous. This system is shown in Fig. 4. The small mass enables the rapid

Fig. 4. Small size furnace with transverse heating system, manufactured by Varian Techtron Ltd, Pty, Melbourne, Australia
Dimensions: length = 9 mm,
O.D. = 5.2 mm,   I.D. = 3 mm,
and weight = 225 mg

heating of small amounts of liquids, and this again is favourable for volatile elements such as lead and cadmium, as higher peaks are obtained by the rapid heating of the small amounts of both sample and sheathing gas within the furnace.

Because of the small dimensions, and the transverse heating system, virtually no condensation and heating problems that exist with the bigger furnaces occur. Because of the fast transient peaks, a very efficient signal analysis system is needed, preferably by a PC. Also the separation of analyte from nonanalyte peaks is a delicate operation: a pyrometer control system is still more necessary here than with the other types of furnace. In Table 1 the combination of a fast signal analysis system combined with the small furnace are compared for the determination of lead and cadmium. For cadmium an improvement in sensitivity compared to the L'vov furnace of about 2, to the GTA 95 of about 7, to the HGA 500 of about 5, and to the two-step Frech furnace [7] of about 3 was obtained. For lead an improvement in sensitivity compared to the L'vov furnace

Table 1. Comparison of the sensitivity different furnaces. CRA 90 furnace pyrometrically controlled; signal analysis by FFT and digital filtering

	Characteristic mass in mg				
Element	L'vov furnace	Varian GTA95	Frech [5] PE HGA 500	Two Step	Herber [12] Varian CRA 90
Cd	0.08	0.3	0.2	0.14	0.041
Pb	2	6.0	13	4.0	0.5

of 4, to the GTA 95 of 12, to the HGA 500 of 26, and to the two-step Frech furnace of a factor 8; clearly a better performance under realistic, routine circumstances.

Platforms, Two-step Atomizers, Different Materials

There have been several publications dealing with improvements, not on the furnace itself, but on the process of phase transition and atomization of the test solution or solid. As there are numerous reviews published on this subject, only a few remarks will be made relating to the general viewpoint outlined earlier that a method should be sample directed.

The optimal choice in chosing a furnace would be to select a furnace for solids, a furnace for volatile elements in liquids, and a furnace for the remaining elements. Regarding the large furnaces for solids and liquids, problems may arise with respect to the unfavourable slow phase transition from solid (after drying/ashing) to gas, due to the slow ramp. A widely accepted improvement for this slow phase transition is the use of the L'vov platform, which enables a better separation between the analyte and nonanalyte peaks for volatile elements such as cadmium [8]. Five possibilities exist, i.e.:

1. Small size furnace for volatile elements in liquids
2. Medium size furnace with platform for volatile elements in liquids (less suited)
3. Medium size furnace without platform for most nonvolatile elements in liquids
4. Large size furnace with platform for solids
5. Large size furnace without platform for solids (less suited).

The optimal combination is limited to three types.

Regarding the two-step atomizers it has been reported, that a considerably improvement is obtained by separating the phase-transition and the atomizing process [7]. However, a drawback of this system is its complexity, which makes it unfavourable for commercial applications.

At the early stages of GF-AAS analyses, experiments were performed with different types of materials. Metals such as tungsten and titanium were used as atomizers [9]. These types of atomizers were hampered by the fact, that no closed atomizer was used, thus giving a very short residence time for the atoms within the optical beam, and low sensitivities. Different types of platforms such as tungsten were used, and many types of carbon. Of these, only the pyrolytical graphite platform is now used widely. Other materials used were coated carbon, e.g. with tantalum carbide [10]. This is an interesting feature because it enables the analyst to choose between the different materials, as in the case of gas chromatography, where many different columns are used depending on the type of test specimen, analyte, and matrix. In fact, this possibility is underdeveloped in GF-AAS, and more extensive research in this area should be carried out.

Pyrometric Temperature Control

The analyte peak time is a function of temperature hence, control of this temperature is of utmost importance. If no feedback system for the temperature is available, the temperature reached at a certain time will be a function of the input power minus the loss of heat due to cooling water, sheathing gas, and radiation losses. Moreover, as the input power is dependent on the resistance of the furnace and electrodes, and resistance is a function of temperature, this leads to complex temperature-time behavior (Fig. 5). It is clear, that it is not easy with such a system to reproduce the same temperature-time diagram, thus it is very unfavourable for samples that have similar physical characteristics. Moreover, as the atomization should start preferably after the steady state temperature has been reached, the control of the temperature is of utmost importance.

In the first setup in our laboratory, the atomization phase of the temperature-time function was pyrometrically controlled by a proportional controller [11]. It turned out, that both the sensitivity and the precision of the determination of lead and cadmium in blood sample improved. In the case of the lead content in blood, the sensitivity, expressed as a slope of the calibration curve, improved by 18%, in the case of cadmium in blood sensitivity improved by 6%. Precision for lead in blood improved by a factor of 1.75 from 11 μg/l to 6.3 μg/l at a level of 150 μg/l, expressed as 95% confidence interval width.

For the determination of lead in blood serum, an improvement in sensitivity of 25% and of precision of a factor of 2 on the very low level of 10 μg/l were obtained. By using the pyrometer control system, extremely low lead levels in serum of about 2 μg/l could be determined routinely.

By controlling the entire temperature program pyrometrically, still more improvements were obtained [12]. In Fig. 6 the temperature profile shows the improvements in obtaining a fast ramp and steady-state as compared with the uncontrolled system of Fig. 5. With this system, both the ramp rate and the platform steady-state temperature could be chosen independently for each temperature phase, i.e. drying, ashing, and atomizing. Figure 7 shows some results for the determination of lead in blood. From this Figure it will be clear, that the highest ramp rate of 1400 K/s gives the highest absorbance signal and

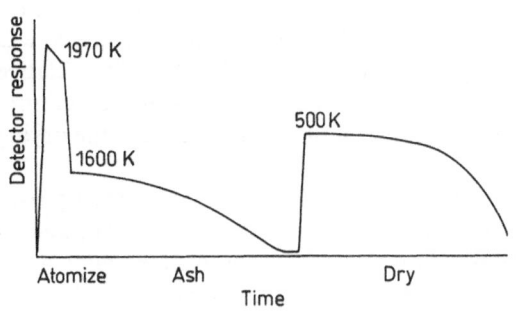

Fig. 5. Temperature diagram of a Varian CRA 90 furnace measured with an optical pyrometer [12]

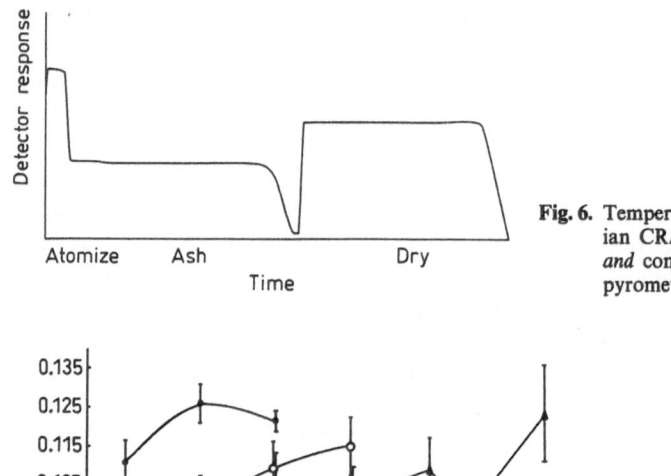

Fig. 6. Temperature diagram of a Varian CRA 90 furnace measured *and* controlled with an optical pyrometer [12]

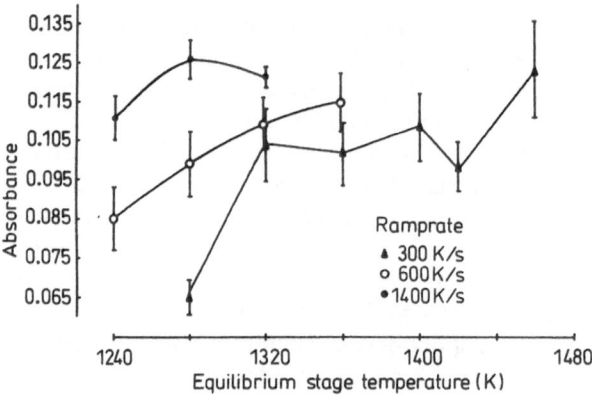

Fig. 7. Performance of a pyrometer feedback system in the determination of lead in blood: dependency of analyte signal upon equilibrium temperature and ramp rate [12]

the best sensitivity. The standard deviation for this ramp rate is at its lowest minimum and the precision is at its maximum for 1400 K/s. It is clear that for this ramp rate an optimal platform temperature is obtained, an optimum that is not seen at the lower ramp rates. Regarding the dry and ashing phases, it was observed that a very narrow optimum regarding the relative standard deviation was occurring as a function of the platform temperature, independently of the ramp rate. The performance of a completely pyrometrically controlled Varian AA6-CRA 63 instrument compared to a commercial instrument, the Varian 1275-CRA 90 combination with pyrometer atomization control, for the determination of cadmium in urine showed that for the former sensitivity was reduced by 19%, but the precision improved from 12.3% to 4.5% relative standard deviation at 10 μg/l. Moreover, the speed of analysis improved from 60 s to 12 s respectively for one firing because of the higher ramps in the drying and ashing stages.

The conclusion is that a pyrometrically controlled heating system improves the precision, sometimes the sensitivity, and permits the unambiguous optimization of both the equilibrium temperature and ramp rate experimental parameters for small furnaces. Moreover, for some circumstances, e.g. for test samples with a low organic content, the speed of analysis may considerably be improved. For larger furnaces especially the precision can be improved; it is expected that optimization here will be easier than in cases without pyrometric control.

Signal Analysis

With the introduction of PC's about 10 years ago, it became feasible to collect a lot of data at low cost. For the graphite furnace-AAS it meant that complete signal analysis became possible. Unfortunately, as already stated, manufacturers were not interested in improving the analytical quality but rather in cosmetic improvements such as windows, colors etc.

Initial improvements regarding signal analysis can be made in the monitoring of the background-time, total signal-time, and temperature-time signals (Fig. 8). The signal from the Varian AA6-CRA 90 instrument is branched directly from the photomultiplier and fed to the Olivetti M 24 personal computer. From the computer an anti-alias filter is directly connected to the A/D converter, enabling the undistorted sampling of the AAS signal. The data are sampled with a frequency of 285 Hz with 6 samples for the 'zero' signal,

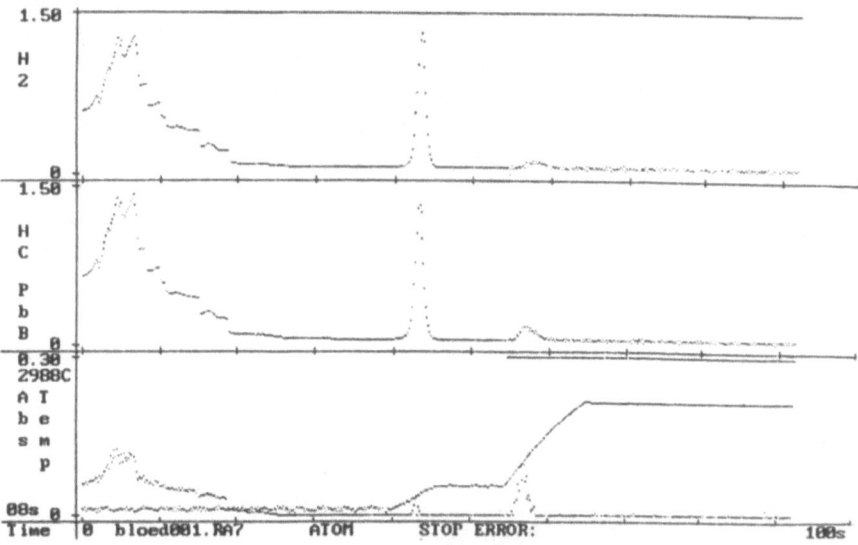

Fig. 8. Monitoring of GF-AAS determination of lead in blood
Upper part: hydrogen continuum source absorption signal
Medium part: lead hollow cathode source total absorption signal
Lower part: corrected absorption signal and temperature signal (the staircase signal). Mountains on left side demonstrate vigorous drying, peak in the middle ashing, and finally peak on the right side atomizing. Atomizing signal time scale (8 s) is enlarged compared to drying and ashing scales

9 samples for both the background and the total signal, i.e. 24 samples per second are taken for the AAS-signal together with 1 sample for the pyrometer signal. The total sample frequency is therefore $25 \times 285 = 7125$ samples/s. The mean of the 6 samples is taken and the same is done for the 9 samples for both the background and total signal. The background signal is monitored in the upper part of the video screen, the total signal in the middle part, and the net signal together with the temperature in the lower part of Fig. 8.

Clearly the extensive drying can be seen in Fig. 8, as well as the ashing peak. In the atomization phase an uncompensated signal can be observed. It is also seen that the analyte peak occurs at the ramp atomization and not at the equilibrium temperature.

A real time monitoring program as shown gives the analyst the possibility to manipulate the different parameters of the determination, as there are the different ramp rates and equilibrium temperatures. Moreover, clear documentation is obtained with all the necessary experimental parameters. Most commercial instruments have a monitor program for the atomization phase, but this is mostly hampered by the fact that the sample frequency is too low for volatile elements, hence, no experimental temperature and no indication about the drying and ashing process is given.

A second improvement regarding signal analysis can be obtained by a digital data acquirement followed by digital filtering techniques. This feature is described more comprehensively in [12]. The principle of data acquisition has been described earlier; subsequent to the data acquisition for the drying, ashing, and atomization phases, a file is made of the atomization phase, only. Figure 9a shows a plot of such a signal. The signal is not disturbed by electronics such as condensators etc. to smooth the signal. This allows a subsequent digital filtering of the signal without initial losses of information.

A digital filtering technique, developed by our laboratory uses Fast Fourier Transformation (FFT). By using FFT, the signal is transformed from the time domain into the frequency domain and gives the power spectrum. The power

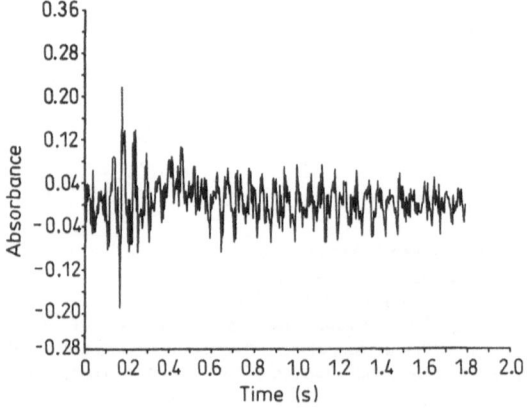

Fig. 9a. Undistorted corrected atomizing signal of cadmium in matrix urine as plotted by computer; data acquisition also by computer [6]

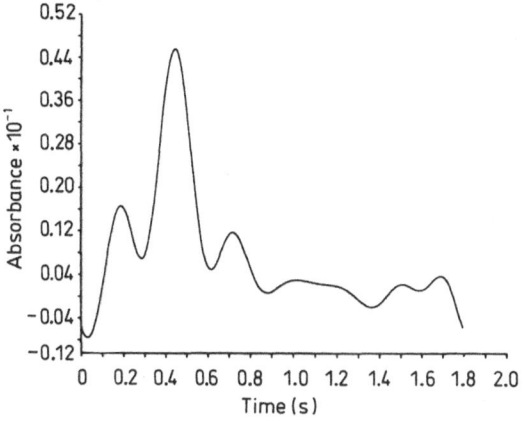

Fig. 9b. Filtered atomizing signal of cadmium in matrix urine as plotted by computer. Fast Fourier Transform back transformation for frequencies below 5 Hz. High noise before analyte peak is due to emission of the furnace [6]

spectrum shows all the frequencies present in the signal. By comparing the frequencies present in a noise signal and those present in the analyte signal, one can conclude that the analyte signal contains only the lower frequencies, e.g. below 5 Hz. Subsequently after FFT, all frequencies above 5 Hz are discarded and the frequencies below 5 Hz are back transformed into the time domain again. Figure 9b shows the result of the back transformation. It shows, that the signal height remains relatively unchanged in contrast to other filter techniques as moving average, envelope etc. is included here. The noise is considerably reduced, hence a far better signal to noise ratio is obtained compared to the non-filtered signal. Compared to electronic filtering, digital filtering is superior by about a factor of 2–3; compared to other digital filtering techniques a gain of 1.5 is obtained, if expressed as signal to noise ratio. In Table 1 the combined effects of a small furnace, pyrometer control and digital filtering is shown. As previously stated, the share of digital data acquisition/filtering is about 2–3.

Another variation of the combined possibilities of pyrometer and digital data acquisition/filtering is shown in Table 2. Here it is seen that under certain circumstances absolute GF-AAS is possible, as reported by L'vov [8]. Another point is also now clear; since GF-AAS is a complicated system, this system should be developed as a whole: e.g. when using a small furnace, fast heating rates are possible which need a fast signal analysis system and a fast power unit.

Background Correction

When GF-AAS was introduced commercially it became clear that especially biological test substances absorbed a lot coatomized matrix material during atomization, leading to wrong absorption signals. Hence a correction system was introduced, based on the absorption of light from a continuum source. The principle is, that nonanalyte atoms and molecules will absorb the light of this

Table 2. Comparison between three peak measurement methods for the determination of cadmium in human urine. Two heating rates are considered

Concentration[a] (µg/l)	Heating rate[b]	Peak time[c] (ms)	Peak height (A)	RSD (%)	Peak area (A s)	RSD (%)	Peak height × time (A ms)	RSD (%)
0.30	L	446	0.046	4.5	2.34	6.1	20.5	1.5
	H	428	0.049		2.55		21.0	
6.3	L	502	0.333	3.9	15.80	3.9	167.2	1.9
	H	488	0.352		16.69		171.8	
22	L	502	0.885	3.9	52.44	5.5	388.8	0.4
	H	488	0.928		48.50		390.7	

[a] Assessed concentration
[b] L: "low", 1200 K s^{-1}; H: "high", 1400 K s^{-1}
[c] Time from start of atomization

continuum source, but not the analyte. Subtracting the continuum source absorption from the hollow cathode atomic light source will give the analyte signal. The first apparatus introduced was equipped with a hydrogen or deuterium continuum lamp and the correction took place sequentially in a subsequent firing. Later on, the correction took place simultaneously within the firing. A drawback of continuum source correction systems is as follows: complicated optical alignment procedures are necessary to ensure correct signals, and correction can only be performed to a limited absorbance. Moreover, the system was not able to correct for molecular band heads.

To overcome these drawbacks, the Zeeman effect correction system was introduced. This is based on the fact, that atom lines in a strong magnetic field may split into more lines, depending on the isotope of interest. Many transverse forms of this Zeeman-effect correction system exist, i.e. with a permanent magnet around the light source (Grün Analysengeräte), with a permanent magnet around the furnace (Hitachi), and with electromagnets around the furnace (Varian, Perkin Elmer 5000 and 3010).

All these systems are advantageous in that corrections can be made for very high background systems and no optical alignment is necessary. The drawback of all these systems is that some form of optical polarization separation system has to be applied to separate the "total" signal from the "background" signal. Some of these systems also correct molecular band heads, i.e. the electromagnet around the furnace system.

Recently, a commercial longitudinal Zeeman-effect correction system was introduced (Perkin Elmer 4100). The advantage compared to the other transverse systems is that no polarization system is needed for separation and thus no loss of sensitivity occurs. Figure 10 shows a theoretical calculation for a

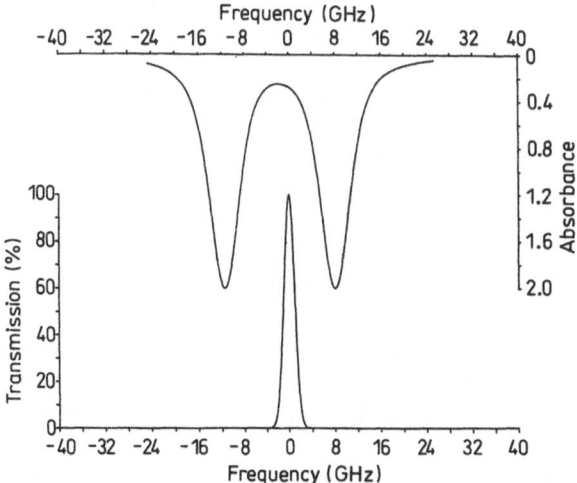

Fig. 10. Model of Voigt absorption and Gauss emission profiles of cadmium. Lamp temperature = 600 K, furnace temperature = 1600 K, λ = 228.8 nm and magnetic field B = 0.7 T

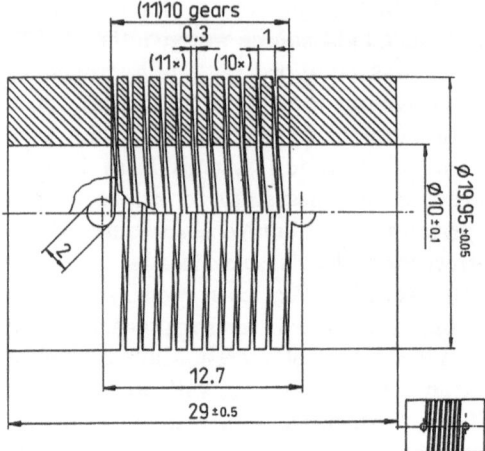

Fig. 11. Coil for Zeeman Electrothermal Atomizer: tungsten. Dimensions are given in mm

longitudinal field system; two absorption peaks appear in contrast to three in transverse fields, thus gaining a factor of 50% in sensitivity. It shows also, that a field strength of 0.7 T is adequate to get a 100% separation in the case of cadmium.

A drawback of the above mentioned longitudinal electromagnetic system is that a relatively high mass is necessary leading to a relatively high induction and hysteresis and to a low frequency modulation of maximum 100 Hz [13]. A practical drawback of having the electromagnet around the furnace systems is that a complicated construction is needed to ensure proper cooling, sheathing gas, and test specimen introduction, which makes servicing and trouble shooting time consuming and expensive.

Hence, a system was introduced without the drawbacks mentioned above, i.e. the Zeeman electrothermal atomizer (ZETA). The core of the system is a tungsten coil, which serves as a furnace for the heating of the test substance and for the magnetic field used for the Zeeman-effect correction system (Fig. 11). The coil is pulsed with a strong current; the peak current is proportional to the magnetic field strength (B). The effective current is related through $I^2 R$ to the energy, and thus to the heat input. Hence, variation of the pulse width may control the heat input. In a pilot experiment, a peak current of I = 420 A was reached for a coil of 25 mm at a temperature of 2300 K and a pulse frequency of 100 Hz. This leads to a B = 0.21 T. Experiments have shown, that temperatures up to 2700 K may be reached with a signal frequency of 300 Hz, at the same magnetic field strength.

The final goal is to reach a magnetic field B = 0.7 T for a coil of 10 mm length. As the signal frequency is rather high, a fast signal analysis system is required (the aforementioned digital sample acquisition/filtering system). However, it is expected that as the weight of the coil to be heated is 20 g, atomization will be slow, and the system will be especially suited for solid test specimens.

References

1. Walsh A (1955) Spectrochim Acta 7: 108
2. Alkemade CTJ, Milatz JMW (1955) J Opt Soc Am 45: 583
3. L'vov BV (1961) Spectrochim Acta 17: 761
4. Kurfürst U (1990) Fres J Anal Chem 337: 248
5. Frech W, Baxter DC, Hütsch B (1986) Anal Chem 58: 1973
6. Herber RFM, Pieters HJ (1988) Spectrochim Acta 43B: 149
7. Frech W, Baxter DC (1987) Fres Z Anal Chem 328: 400
8. L'vov BV (1979) Spectrochim Acta 33B: 153
9. Ortner HM, Sychra V (1985) Book of Abstracts Spectrosc Internat vol 2, p 340, Garmisch Partenkirchen
10. Almeide MC, Seitz WR (1985) Appl Spectrosc 40: 4
11. Deyck W van, Roelofsen AM, Pieters HJ, Herber RFM (1983) Spectrochim Acta 38B: 791
12. Herber RFM, Pieters HJ, Roelofsen AM, Deyck W van (1984) Spectrochim Acta 39B: 397
13. Loos-Vollebregt MTC, Galan L de (1988) Spectrochim Acta 43B: 1147

6.5 Occurrence and Behaviour of Mercury and Methylmercury in the Aquatic and Terrestrial Environment

S. Padberg, K. May

The distribution of mercury and methylmercury in aquatic and terrestrial ecosystems and in the human food chain was studied. Using a modified cold vapour AAS for the determination of mercury very low concentrations could be analyzed in different environmental and biological matrices. Methylmercury was separated by ion exchanger after hydrochloride extraction or by water vapour distillation with H_2SO_4/NaCl. It was possible to investigate the transport and the transformation mechanisms in the aquatic and the terrestrial environment. Thus, information about the complex interactions between the different compartments regarding the distribution of mercury in the environment and its possible impact on humans by the uptake of food was obtained.

1 Introduction

Mercury is of particular significance in environmental pollution due to the toxicity of its organic compounds and their characteristic of accumulating in the food chain.

Elementary mercury ($Hg°$) escapes into the atmosphere at low temperature (1.7×10^{-3} mbar at $24\,°C$) due to its relatively high vapor pressure. Of the inorganic Hg(I) and Hg(II) salts, the former are significantly more reactive and readily form complexes with organic ligands, particularly with sulfhydryl groups. However, in contrast to $HgCl_2$, Hg_2Cl_2 is not easily dissolved and thus less toxic. Organomercury compounds have various chemical structures. The organic cations form salts with inorganic and organic acids and thus react equally rapidly with ligands, in particular with sulfhydryl groups. Due to their solubility in fat, they easily penetrate biological membranes and accumulate in the cell. Methylmercury (MeHg) is of greatest significance in environmental pollution since it is the biologically most resistant alkyl mercury compound [1]. The methylated form of Hg(II) (CH_3HgX) consists of a covalent bond with an organic radical ($R = CH_3$) and an electrovalent bond with an inorganic anion (X), frequently chloride ion ($R–Hg–X = CH_3HgCl$).

As long ago as the 1930s, Alfred Stock was the first analytical chemist to work with environmentally relevant materials with respect to their mercury content [2]. From his investigations, it is known that mercury is almost ubiquitously present in the environment, even if it occurs naturally in extremely small quantities [3, 4].

Specimen Banking
Rossbach/Schladot/Ostapczuk (Eds.)
© Springer-Verlag Berlin Heidelberg 1992

The occurrence of these trace elements in all compartments is primarily attributed to the geochemical distribution, particularly to Hg deposits close to the surface [3, 5] and to the physico-chemical properties of mercury and its compounds [1, 6].

Due to weathering processes and volcanic exhalations mercury is released at relatively high vapor pressure and enters the natural cycle as elemental, volatile Hg via the air [7] and primarily as divalent Hg salts via water [1, 6].

The geochemical cycle is subjected to the interactions and exchange processes of the various compartments and determines the natural Hg concentrations – the geochemical background – in the environmentally relevant matrices (Table 1).

However, anthropogenic activities have altered the geochemical global cycles during the past few decades [9]. Local contaminations result from emissions due to the exploitation of deposits [10–14], and the industrial and agricultural application of mercury and its compounds [3].

Catastrophic toxic accidents in the fifties–seventies with Hg emissions – on occasion in the form of either organic mercury compounds or methylmercury (Japan, Iraq, Sweden) – leading to serious consequences (Minamata disease) [1, 15, 16] gave rise to numerous studies on environmental pollution by mercury and its compounds in natural systems.

It became apparent that the ecological, physiological and toxicological effects depend on the form or the species [1, 9]. The uptake of methylmercury through food results in the destruction of brain cells via blood transport due to the sulfhydryl group affinity of the amino acids. Furthermore, this Hg compound causes irreversible genetic damage [17, 18].

With only a slow metabolization, methylmercury is bioaccumulated particularly in fish so that methylmercury content in muscle tissue amounts to more than 80% of the total Hg content [1, 18, 19]. The toxicological significance of even minimal concentrations of methylmercury in the environment is intensified by its characteristic of accumulating in the food chain [20–23].

Anthropogenic emissions are responsible for methylmercury contaminations in Japan, Iraq, and Sweden. However, the occurrence of methylmercury in fish

Table 1. Natural mercury contents [3, 4, 8]

Occurence	Hg content
Earth's crust	0.08 mg/kg
Clay minerals	0.01–0.2 mg/kg
European soils	0.04 mg/kg
Soils in mining area	< 10.0 mg/kg
Coal/fossil oil	ca. 1.0 mg/kg
Volcanic exhalations	< 0.02 mg/m^3
Atmosphere	< 0.002–0.02 μg/m^3
Precipitation	< 0.005–0.03 μg/L
Spring water	< 0.001–0.05 μg/L
Oceans	< 0.001 μg/L
Plants	0.05 mg/kg

from remote regions with no known anthropogenic sources of methylmercury raise questions as to its origin and the possible formation mechanisms (methylation) in nature, and their conditions in the environment [21, 24–27]. Nevertheless, studies have concentrated on the mercury cycle in aquatic ecosystems. Less attention has been paid to Hg cycles in the atmosphere and in terrestrial ecosystems, the input of methylated mercury by precipitation, and the drainage of terrestrial systems into aquatic systems. This was due to analytical problems, primarily in determining methylmercury in atmospheric matrices and in the inhomogeneous soil matrix, where it is frequently present only in extremely low concentrations, but which are toxicologically highly effective [28–31].

Mercury analysis in various environmental matrices has been carried out since 1976 at the Institute of Applied Physical Chemistry at the Research Centre Jülich (KFA). Due to the modification of an effective cold vapor AAS method and the development of reproducible separation procedures for MeHg determination [32–38], mercury speciation in aquatic and terrestrial systems is documented within the framework of the German Environmental Specimen Bank (ESB) [34, 39, 40]. Furthermore, experiments are continuously being implemented to provide contributions towards solving questions regarding the complex interactions in aquatic and terrestrial systems [37, 41–45].

2 Material and Methods

Precise working methods are required for a study of the behaviour of mercury and methylmercury present in the environment in low concentrations. This does not only involve the application of effective detection and reproducible analytical methods: representative sampling should also be organized meaningfully, and if the specimens are to be preserved and prepared care should be taken to avoid both contamination and also losses.

During the pilot phase of the Environmental Specimen Bank, 1980–1985, numerous investigations were carried out with respect to this problem so that in permanent operation contamination-free sampling and storage is now ensured [40]. Contamination or loss due to microbial conversions and the resulting volatilizations is prevented by acidification of the samples (liquid matrices) and by storage below $-30\,°C$ (solid matrices) in glass containers in the absence of light.

2.1 Measuring Method

The cold vapor method used is based on Hatch and Ott's (1968) principle. After extraction and separation steps for Hg speciation, and after oxidation of the Hg compounds in acid solution (decomposition process), ionic Hg is reduced by adding tin(II) chloride (10% $SnCl_2$ in 20% H_2SO_4) to elementary mercury [46].

In the modified procedure, the detection sensitivity is significantly increased by preconcentration of Hg vapor on gold wool (amalgamation). Moreover, the mercury vapor is purified in sodium hydroxide solution and deionized water before amalgamation in order to prevent contamination and matrix effects by the acids [32, 35]. A drying cell (silica gel) connected in series ensures the stability of the gold wire [34].

The absolute detection limit of this modified measuring procedures is in the region of 0.0005 ng [34]. Hg and MeHg concentrations of 0.1 μg/kg in solid samples and 0.2 ng/l in aqueous matrices can thus be determined. Experiments by May et al. [47], show that the yield is in the region of 98% and the variation coefficient for contents >200 μg/kg is about 2% and even for contents >1 μg/kg it does not reach more than 10%.

2.2 Decomposition of Mercury and Mercury Species

Cold vapor ASS requires mercury to be present in the form of ionogenic mercury in solution (Hg^{2+}) [46]. The total mercury content or else the various extracted and separated mercury compounds must therefore be converted into the ionogenic compounds by a decomposition process.

Total mercury analysis is carried out for solid materials with a sample weight of 1–2 g after wet decomposition at low pressure in closed quartz vessels with an $HNO_3/HClO_4$ mixture (10 ml, 1:1) at 150 °C [48]. During this decomposition (about 4 bar) the mercury compounds are oxidized to Hg^{2+} as required for determination by the cold vapor AAS (see above). Total mercury is determined in unfiltered and filtered (0.45 μm filter) liquid matrices (freshwater, seawater and rainwater) after UV irradiation in solutions acidified with HCl and/or HNO_3 (to approx. 1%) by the modified cold vapor AAS [42, 49].

In accordance with the gas chromatography method for MeHg determination developed by Westöö [50], the separation of methylmercury from inorganic ionic mercury compounds is based on extraction with hydrochloric acid. Up to 2 g of solid material was extracted twice with 10 ml 6 M HCl for this purpose. The methylmercury present in the hydrochloric acid extract of the solid materials (20 ml) or in water (500 ml) acidified with 6 M HCl was quantitatively separated from the inorganic ionic mercury compounds with an anion exchanger (Dowex 1 × 8, Cl form, 100–200 mesh). MeHgCl passes through the exchanger, while the inorganic ionic Hg compounds are adsorbed on the column as chlorocomplexes ($HgCl_4^{2-}$) [35, 47]. The eluate was directly analyzed by CVAAS in the absence of light for its Hg content in order to recognize the contamination from inorganic Hg compounds. MeHgCl present in the ion exchanger eluate was decomposed to Hg^{2+} with UV irradiation (5 h) in the quartz vessel [51–53] and was thus made available for measurements with cold vapor AAS.

The methylmercury concentration analyzed by the ion exchanger method was checked by a water vapor distillation. For this purpose, 750 ml of the acidified aqueous samples was distilled to 500 ml and 10 ml HCl and 2 ml NH_3 were added. The methylmercury exclusively present in the distillate as a chloride was converted into the ionogenic form by UV irradiation.

MeHg concentrations up to 0.2 μg/kg were determined in soils with a yield of 90–95% by the following distillation method and verified by a modified separation procedure and subsequent determination by gas chromatography [34, 36–38]. The fresh soil material as taken from the field (0.2 g– 5 g) was directly distilled without extraction by the addition of 8 ml H_2O, 1 ml H_2SO_4 (1:1), and 1 ml NaCl (10%) at 150 °C and an air flow rate of 14 l/h [36]. The MeHg yield was increased and stabilized by the separation of the remaining ionic Hg compounds and matrix-related components from the distillate with an ion exchanger. Therefore the distillate (9 ml) was subsequently passed through the ion exchanger column after the addition of 1 ml of concentrated HCl and then eluted with a further 10 ml 6 M HCl [37, 38]. The eluate (20 ml) was measured immediately by CVAAS in the absence of light to obtain the Hg content. MeHgCl was subsequently decomposed by UV light (1–3 h) and the content determined by CVAAS.

3 Mercury in Aquatic Systems

Aquatic systems represent a reservoir for all emissions of Hg, whether as wastewater or as atmospheric input by precipitation or air exchange. Accumulation and metabolic processes take place in the ecosystem water-sediment-life.

The sediment takes on a particular role since it represents a store for the pollutants. Due to the temporary fluctuations and different flow rates, the water body itself only reflects an indistinct picture of the degree of the load. Since the heavy metals–hence also the Hg compounds–are mostly bound to the suspended matter they finally accumulate in the sediment [54–58]. The biological and physico-chemical conditions of the sediment are not only responsible for the mineralization and accumulation of mercury [54, 59–62]. Whirling up and chemical metabolic processes at the water-sediment interface may also remobilize mercury [63], i.e. it becomes available for chemical transformation processes such as methylation or demethylation [19, 64–69] and for uptake into the food chain [20, 70, 71].

These processes taking place in the aquatic ecosystems were investigated in various compartments of the limnetic (Table 2), fluvial (Table 3), and marine systems (Tables 4, 5). The two ponds investigated in 1986 [44, 72] differ considerably with respect to their total Hg content in the sediment and only slightly with regard to the concentration in the water body, where the content in pond A is greater than in pond B (Table 2). However, this is only true for the total mercury content: MeHg contents in water and suspended matter at low average values of about 1 ng/l in the two bodies of water only display an increase of 0.2 ng/l–0.3 ng/l in pond A as opposed to pond B. The ponds also hardly differ with respect to their MeHg load in the sediments.

Above 80% of the Hg content in the water body is bound onto individual particles in both ponds. Due to complex reciprocal actions in the sediment and water body, the total Hg content in water is increased in pond A in comparison

Table 2. Mercury concentrations in two ponds in the Sauerland [44, 72]

| | Pond A | | Pond B | |
	Total Mercury (mean ± S.D.) (range)	Methylmercury (mean ± S.D.) (range)	Total Mercury (mean ± S.D.) (range)	Methylmercury (mean ± S.D.) (range)
Water[a] (ng/l)	13.29 ± 4.57 6.33–21.68	1.05 ± 1.31 0.10–5.15 7.9% of tot. Hg	3.94 ± 2.77 1.89–13.91	0.81 ± 0.65 0.16–2.51 20.5% of tot. Hg
Suspended matter (ng/l)	11.68 ± 3.89 4.71–18.11 87% of tot. Hg in water	1.16 ± 1.21 0.08–4.95 9.9% of tot. Hg	6.19 ± 4.14 2.10–15.31	0.83 ± 0.71 2.10–2.48 13.4% of tot. Hg
Sediment (ng/g, cal.dw)	1078.39 ± 137.7 842.57–1301.23	6.72 ± 1.77 3.21–9.28 0.6% of tot. Hg	132.80 ± 48.4 52.04–226.20	5.59 ± 3.01 2.22–12.45 4.3% of tot. Hg
Fish[b] (ng/g, cal.dw)	85.56 ± 1.76 43.28–177.53	77.92 ± 2.45 40.58–152.98 91.6% of tot. Hg	80.17 ± 3.20 35.03–113.67	73.18 ± 3.29 32.22–101.49 90.3% of tot. Hg

[a] Unfiltered sample
[b] Trout, muscle

to pond B. However, no increase corresponding to that in the sediment (factor 8) can be seen but rather merely an increase by a factor of 2–3.5. With respect to Hg content in fish (35–177 μg/kg), the ponds differ only slightly.

The mercury distribution within the pond system between dissolved and suspended bound phase results from accumulation, remobilization, exchange, and transformation processes. It is thus controlled in a different manner by the autochthonous conditions of the respective waters in ponds A and B.

The different total Hg load in ponds A and B must, however, be attributed to the input of mercury. This input can result both from atmospheric emissions as well as from fluvial transport. It can, however, be assumed that the atmospheric emission of Hg is the same for both ponds since due to the geographical position the climatic conditions are identical. The different mercury content in the ponds therefore results from the fluvial input from the drainage area. Mercury enters pond A from a Pb–Zn deposit in its drainage area. Hg contaminations in the soil material were determined in the region of the deposit due to the occurrence of mercury as an accompanying mineral to zink blende (ZnS) [10]. A further source for transport of Hg with the surface water is represented by the spoil tip from the working of ore deposits.

The significance of the fluvial transport of mercury similarly becomes apparent in a study of the Hg load of the River Elbe [73]. With respect to its Hg load, the Elbe is one of the most heavily polluted rivers in the world with Hg concentrations for suspended matter of about 8–60 mg/kg [74] and about 10–70 mg/kg for sediments [75]. The transport behaviour depends on the interactions in the fluvial system which influence the phase distribution. Comparative determinations of the total Hg content in unfiltered river water in the region between Schnackenburg and Hamburg, between Hamburg and the mouth of the estuary (lower Elbe, tidal Elbe), and in the estuary region also confirm the high fraction of mercury bound to particles in the water complex (>80%). As a function of seasonal fluctuations and the fluctuating salt content, particularly in the tidal region, a fluctuation of the dissolved Hg fraction of between 3 and 12% was observed [76]. Seasonal fluctuations thus do not only bring about a different flow rate but also a different sedimentation rate. Sedimentation and remobilization then finally cause the transport which in the course of the river becomes less apparent in the water complex than in the sediments.

Current data on the Hg load of the Elbe in the region between Postelwitz (FRG/CSFR) and Schnackenburg (Table 3) similarly confirm that the Hg concentrations in the water complex are not representative of the course and degree of pollution. With total Hg contents of 119–300 ng/l in the unfiltered river water, the Hg concentrations in the sediment are higher by a factor of about 10^4–3×10^4. The similarly high fraction of mercury bound in a particulate form (>80%) causes rapid sedimentation.

The influence of the fluvial input can be clearly seen from the Hg concentrations in areas of the North Sea and Baltic Sea close to the coast (Table 4).

Table 3. Mercury concentrations in water and sediment of the Elbe: Postelwitz to Schnackenburg, Jan. 1991 [77]

Sampling Location	Water[a] Total mercury (ng/l)	Total mercury (µg/g)	Sediment (ng/g)	Methylmercury (% of tot. Hg)
Postelwitz	119.1 ± 26.2	1.13 ± 0.01	6.15	0.5
Dresden	133.9 ± 14.3	1.53 ± 0.07	21.05	1.4
Dressau	142.9 ± 57.3	n.d.	n.d.	n.d.
Barby	295.5 ± 38.2	4.94 ± 0.51	22.73	0.5
Magdeburg	246.0 ± 6.7	4.12 ± 0.01	14.69	0.4
Havelberg	300.3 ± 51.1	2.50 ± 0.01	14.40	0.6
Cumlosen	214.3 ± 7.2	6.08 ± 0.08	7.31	0.1

[a] Unfiltered sample
n.d. not determined

Table 4. Mercury concentrations in water (unfiltered) from areas of the North Sea and Baltic Sea close to the coast [78]

Sampling area	Mean (ng/l)	Range (ng/l)	N
North Sea – close to beach (nonpolluted)	9.3	4.6–15.3	10
North Sea – close to beach (polluted)	46.4	11.5–50	10
Baltic Sea – close to beach (nonpolluted)	2.4	0.4– 3.7	10
Baltic Sea – close to beach (polluted)	39.0	2.5–130	10
Baltic Sea – open sea	0.6	0.2– 2.2	8

3.1 Methylmercury in Aquatic Systems

Different chemical conditions in the sediments lead to different accumulation or remobilization processes and thus to a different provision for possible transformation processes. A transformation process which may take place in the aquatic system is mercury methylation and demethylation.

The MeHg concentrations in the Elbe sediment (Table 3) fluctuate between 6 μg/kg and 23 μg/kg. With 0.1–1.4% of the detected total Hg content they do not behave in proportion to the total Hg supply. This phenomenon was also observed in the sediments of the ponds investigated in Table 2. As already mentioned above, the MeHg contents in the two ponds fluctuate between 2 μg/kg and 12 μg/kg, and with an average of approx. 6 μg/kg they represent 0.6% of the total Hg content in pond A with a total Hg content of approx. 1000 μg/kg and 4% in pond B with a total Hg content of approx. 130 μg/kg. On the basis of this MeHg content, which does not increase proportionally to the total Hg content, the sediments from ponds A and B loaded to a different extent with Hg were characterized in detail with respect to their chemism in order to obtain indications of any possibly different transformation processes.

Factors such as increased sulfur content in the case of redox conditions below − 145 mV with slightly increased pH lead, as shown by Brümmer [79] in laboratory experiments, to an immobilization of the mercury. These factors are more pronounced in pond A than pond B. In the same way, the increased fraction of organic substances in the more heavily polluted pond system has more significance for the binding of mercury in comparison to clay minerals. A greater immobility of the mercury in this pond system leads to a lower availability for possible methylation processes and thus explains the, in comparison to pond B, lower methylation rate (proportional MeHg content in comparison to the total Hg supply). Moreover – according to Craig's theory

[80, 81] – the increased sulfur content in pond A would rather lead to a formation of volatile dimethylmercury.

The possibility, frequently described in the literature, of the formation of methylmercury in aquatic systems [65, 66, 82, 83] was similarly detected in laboratory experiments under controlled conditions [41, 84]. An increase of MeHg in sediment was observed in aquarium experiments – equipped with the sediment, plants, and snails of a natural limnetic ecotope – after the addition of HgCl$_2$. The formation of MeHg in the sediment was confirmed in a replicate of the experiment with radioactively labelled Me ^{203}Hg.

The occurrence of MeHg in natural aquatic systems may be regarded as the result of a process of equilibrium being established between methylation in the sediment and simultaneous demethylation processes. However, at the same time attention should be paid to the input of MeHg into the aquatic system from the atmosphere and the drainage of terrestrial systems (cf. Sect. 4). The MeHg concentrations in the water complex result from the fraction formed and mobilized in the sediment and the fraction from outside input. The MeHg concentrations in the biotic compartment of the aquatic system do not result from a methylation in plants or animals but rather from the uptake from sediment and water body and the accumulation effect in the food chain.

3.2 Bioaccumulation Studies in Aquatic Systems

In accordance with the environmental conditions and in accordance with the food uptake, mercury compounds enter into the trophic stages of the food chain in various ways. Within the framework of a permanent investigation series by the ESB, samples from various materials of selected representative sites in the area of the North Sea (mud flats, see Schladot JD and Backhaus F in this volume), close to the coast can be continuously studied (Table 5) [40, 85]. If accompanying studies are included, clear relations can be perceived between the pollution of the links in the food chain of the marine coastal areas and the Hg pollution of the river systems flowing into them [40].

Whereas algae mainly take up mercury in a dissolved form, mussels are capable of filtering mercury and methylmercury bound to particles out of the water body [86]. This is why a higher fraction of methylmercury is to be found in mussels than in algae. Dissolved Hg ions and complexes are taken up by plankton and microorganisms through adsorption and diffusion [87]. Aquatic plants incorporate dissolved Hg compounds into their metabolism by assimilation and ingestion via the roots from water and pore water [61]. Benthic and pelagian aquatic fauna feed on plankton containing Hg, organic suspended and sedimented particles, microorganisms and plant parts. The chemico-physical conditions of the water (temperature, pH, salinity, hardness, oxygen content, content of organic substances) do not only decisively influence the provision of mercury for the organisms but also the metabolic processes of the organisms and thus, the uptake, effect, and toxicity of mercury [61].

Table 5. Mercury concentrations in water (unfiltered), brown algae (*Fuscus vesiculosus*), common mussels (*Mytilus edulis*) and herring gull eggs (*Larus argentatus*) [77]

Material	Sampling period	Sampling area			
		Eckwarderhörne / Trischen		List/Sylt / Mellum	
	Number of samples	Total Mercury (Mean ± S.D.) (Range)	Methylmercury (Range) (% of tot. Hg)	Total Mercury (Mean ± S.D.) (Range)	Methylmercury (Range) (% of tot. Hg)
Water[a]	1989 n = 16	60.20 ± 53.37 8.8–153.75	< 0.2–1.00 3.95–103.5	29.31 ± 32.41	< 0.2–1.00
	1990 n = 16	56.03 ± 52.18 8.4–133.9	< 0.2–1.68 3.62–44.2	15.24 ± 15.89	< 0.2–1.43
Brown algae	1989 n = 12	23.00 ± 2.27 20.0–26.45	< 0.2–0.8 < 3.5%	8.74 ± 1.24 7.42–10.9	< 0.2–0.65 < 6%
Common mussel	Feb. 1990 n = 12	34.3 ± 9.8 26.6 ± 59.1	1.9–7.86 13.6%	38.5 ± 7.4 21.1–49.1	7.45–12.16 25%
		Trischen		Mellum	
Herring gull eggs	1989 n = 5	0.82 ± 0.33 0.52–1.38	0.57 ± 0.13 70%	0.24 ± 0.10 0.11–0.37	0.20 ± 0.08 84%
	1900 n = 15	1.04 ± 0.25 0.63–1.65	83%	0.27 ± 0.12 0.12– 0.55	82%

[a] unfiltered sample

The greatest bioaccumulation is to be observed in predatory fish representing the final link in the aquatic food chain. However, the uptake of mercury from water via the respiratory organs can also be determined here [88]. Mercury concentrations in bream (Abramis brama, 500–2000 µg/kg) and smelt (Osmerus eperlamis, 700–2000 µg/kg) from the heavily polluted Elbe [76] are significantly above the contents in trout (Salmo trutta, 30–180 µg/kg) of the ponds studied in the Sauerland [44]. Over 90% of the Hg concentrations detected in fish are present as methylmercury (see Table 2). In the long term, the MeHg is accumulated in muscle tissue whereas inorganic mercury, when taken up, is excreted again in the metabolic process of the liver. Although the acute Hg pollution is reflected by the concentration in liver, nevertheless the accumulation and toxic effectiveness of MeHg becomes apparent from concentration of muscle tissue [20].

The Hg concentrations in herring gull eggs (*Larus argentatus*) from the bird reserves of Mellum and Trischen provide a particularly good reflection of the burden of the Weser (Mellum) and Elbe estuary (Trischen) as the values of a sedentary final link in the aquatic food chain [40]. The extremely toxic methylmercury also represents the greatest fraction of total Hg concentration in this case as well.

4 Mercury in Terrestrial Systems

In the research project 'Mercury – occurrence and turnover of mercury in the environment' initiated in Sweden in 1984 by the National Swedish Environmental Protection Board, the increasing MeHg contamination of fish in Swedish lakes, in contrast to the decreasing MeHg emissions, is being studied for the first time from the view point of an interconnected Hg cycle between the atmosphere, aquatic, and terrestrial system [31]. The significance of soils becomes clear with respect to solubility and binding as a function of various soil properties [89]; hence, for the fixation of the mercury or its provision for the conversion processes and transport in aquatic systems and for uptake into the food chain.

It is particularly this heterogeneous matrix consisting of solid components such as various organic and organomineral substances, clay minerals, iron, aluminum, and manganese oxides, pore volumes of various sizes filled with gases and water with dissolved substances, and those bound in particles which is responsible for different soil reactions and redox properties. The binding mechanisms and remobilization conditions vary as a function of the soil composition, the surface reaction, the external and internal specific surfaces with the various load densities [90]. Furthermore, the composition varies in accordance with the pedogenesis, which is dependent on the geological and climatic conditions, increasing soil depth, changes in vegetation, and the geographic site.

It is therefore of particular interest to clarify the extent to which toxic effects of mercury are determined by the adsorption behavior and the degree to which soil-specific properties alter the transformation, metabolization and bioavailability.

Samples from the terrestrial system are being studied at various sites in the FRG (Berchtesgaden, Bavarian Forest, Saarbrücken Bornhöved) within the framework of the ESB programme [40]. Detailed studies were undertaken at a location in the Sauerland (Hunau) in order to clarify questions concerning the input of mercury and its behavior in soil [37]. An accumulation of mercury in the humus-rich topsoil (Ah horizon) and a decrease in concentration with increasing soil depth is apparent for all soil profiles from all sites (Table 6).

Studies at the location Hunau revealed that the accumulation in the topsoil is caused by the strong sorption of mercury onto the humic substances due to complexation [91]. Since the binding capacity of the organic fractions is particularly high in comparison to inorganic components under acidic conditions [92] an absorptive binding of mercury to the clay minerals hardly becomes effective at all. Determination of the binding types after selective extraction [93] confirms Hg fixation in the top soil due to predominant binding to the humin and fulvin fractions and the continuing dominant complexation onto fulvins in the subsoil.

The distribution of binding types in the vertical profile similarly provides indications of the mobility or transport of mercury in the soil complex. Eh and pH conditions determine the mobility of the water-soluble fulvins frequently bound to Fe and Al oxides, and thus the transport of the mercury complexed onto this organic fraction.

At the Hunau site, the input of mercury and methylmercury into the soil complex is derived primarily from atmospheric deposition and from the input of vegetation constituents. The respective deposition rate becomes apparent from the different loads in the layer profile (O horizon) of the profiles at different sites (Table 6).

No formation of methylmercury proportionate to the total mercury content was established at any of the investigated sites (Table 6). A transformation of mercury in the sense of a methylation seems improbable under the given Eh and pH conditions at the Hunau site investigated in detail. The distribution in the profile is similarly determined by the distribution or the Eh/pH-dependent mobility of the organic substance. However, on the basis of this exemplary observation, the possibility of Hg methylation taking place in soil as a function of the supply of humic substances [43, 94, 95] and the bacterial activity as described in the literature on the basis of laboratory experiments [96–98] cannot be completely ruled out. Such an observation is undoubtedly easier to make in the laboratory under controlled boundary conditions since an equilibrium is established more rapidly under natural conditions due to the degradation processes taking place simultaneously.

The mercury present in the soil solution as a metallo-organic complex also forms the greatest fraction of mercury available to plants in both profiles from the Hunau site. An accumulation of mercury and methylmercury was observed in spruce needles of different ages with increasing exposure time or deposition, and with increasing metabolic activity or ingestion [37].

After separating the needle surface, and thus the particles deposited on it [99], more than 90% of the mercury was detected in the needles both in the

Table 6. Mercury concentrations in various soil profiles [37, 77]

Sampling area/period		Soil horizon	Total Mercury (Mean ± S.D.)		Methylmercury (Mean ± S.D.)	(% of tot. Hg)
A	2.89	Of	100.82	± 20.88	0.82 ± 0.18	0.8
		Oh/Ah	124.31	± 33.81	0.5 ± 0.06	0.4
		Bv	11.81	± 2.53	0.42 ± 0.02	3.5
	8.89	Of	135	± 10		
		Ah	52	± 2.8		
		Bv	18.8	± 0.3		
		C	7.52	± 0.26		
B	3.89	Ol	115.5	± 0.7		
		Ah	356	± 24		
		Go	30.8	± 1.1		
	8.89	Of/Aw	80.12	± 1.85	2.82 ± 0.3	3.5
		Ap	77.37	± 2.24	2.53 ± 0.09	3.3
		Ap/Bv	83.33	± 6.26	2.58 ± 0.49	3.1
		Bv	38.24	± 0.81	0.82 ± 0.06	2.1
		Ap	70.55	± 1.4	3.69 ± 0.05	5.2
		Bv	45.16	± 1.04	1.46 ± 0.21	3.2
		Bv/Cv	26.19	± 0.56	1.24	4.7

Location	Date	Horizon			
C	9.89	Of	105 ± 6		
		Ah	110 ± 6		
		Bv	55.4 ± 1.2		
		Of	87.2 ± 1.5		
		Ah	124 ± 5		
		Bv	69 ± 1.3		
		Cv	6.1 ± 0.4		
D	3.89	O	60.0 ± 6		
		MI	4.5 ± 0.9		
		MI-III	4.45 ± 0.8		
E	6.89	Of-h	66.07 ± 1.80	0.61 ± 0.24	0.9
		Ah	196.44 ± 3.39	0.38 ± 0.14	0.2
		Bv1	68.24 ± 7.06	0.37 ± 0.05	0.5
		Bv2	50.85 ± 5.42	0.06 ± 0.07	0.1
		Cv	34.82 ± 1.90	0.27 ± 0.11	0.8
		L,Of-h	213.07 ± 6.32	1.40 ± 0.13	0.6
		Ah	226.12 ± 29.74	0.46 ± 0.11	0.2
		Bv1	80.06 ± 6.25	0.19 ± 0.06	0.2
		Bv2	58.29 ± 12.95	0.19 ± 0.15	0.3
		Cv	47.16 ± 1.38	0.04 ± 0.06	0.1

Locations: A – Bornhöved; B – Saarbrücken; C – Bayerischer Wald; D – Berchtesgaden; E – Hunau (Sauerland)

Table 7. Mercury concentrations in leaf organs at various sites [37, 77]

Sampling area/ period	Spruce twigs	Beach leaves	Popular leaves
Bornhöved[a]			
2.89	12.9 ng/g (n = 15) 10.2–15.75 ng/g		
8.89		22.45 ng/g (n = 16) 18.7–28.8 ng/g	
Saarbrücken[a]			
3.89	17.7 ng/g (n = 13) 13.4–20.55 ng/g		
8.89			18.20 ng/g (n = 10) 11.6–22.05 ng
Bayerischer Wald[a]			
3.89		25.28 ng/g (n = 5) 19.35–28.65 ng/g	
Berchtesgaden[a]			
3.89	10.5 ng/g (n = 15) 7.55–16.2 ng/g		
Hunau (Sauerland)			
6.89	36.09 ng/g (n = 10) 18.26–55.58 ng/g	9.51 ng/g (n = 4) 8.27–10.76 ng/g	

[a] Sample material from Environmental Specimen Bank

young spruce shoots and also in those several years old. It was, however, determined that the concentration in leaf organs of the plants primarily resulted from intake from the atmosphere, whereas uptake via the root system from the soil seems to be insignificant as far as accumulation in the leaf organs is concerned [100]. The vegetation takes on the role of a filter with respect to atmospheric Hg pollution. The particles deposited on the leaves by atmospheric wet and dry deposition may enter into the leaf through the cuticle by means of diffusion. Furthermore, elementary Hg° is taken up from the atmosphere by the stomata, and then stored and accumulated in the leaf organ [101, 102]. The different Hg concentrations in leaf organs at various sites (Table 7) and various ages thus reflect the local Hg pollution of the atmosphere.

In the case of Hg pollution of soil, the vegetation is a source of an additional input. With respect to the Hg content in the wet deposition directly reaching the grassland, the vegetation merely represents an interim store. Nevertheless, it additionally takes up particles from dry deposition and gaseous Hg compounds from the atmosphere which would not reach the grassland without precipitation events. Mercury removed from the soil, even if to a lesser extent, mainly remains in the roots. The mercury filtered out from the atmosphere and concentrated by

Table 8. Mercury concentration in the terrestrial food chain [77]

Material	Total Mercury (mean or range–ng/g)	(range–ng/g)	Methylmercury (% of tot. Hg)
Rye grass[a] (*Lolium multiflorun*) – leaves	1.5–4.7	< 0.2	10%
Poplar leaves[a] (*Populus nigra italica*)	10–25	< 0.2	0.5%
Spruce twigs[a] (*Picea abies*)	6.6–60	< 1	< 2%
Earthworms[a] (*Lumbricidae*) – total specimen	72–127	4–12	10%
Beetle[a] (*Carabus auratus*)	67	41	61%
Bird feathers – Blackbird (*Turdus merula*)	1400	812	58%
– Goshawk (*Accipiter gentilis*)	2800	2300	82%
Bird maw – Blackbird (*Turdus merula*)	79–177	7–47	
Fox (*Vulpes vulpes*) – muscle	15–110	7.5–98	80%

[a] Sample material from Environmental Specimen Bank

fallen leaves and dead root constituents and only briefly removed from the soil is fed into the soil.

One aspect concerning the Hg cycle in the terrestrial system is the mercury and methylmercury pollution of the atmosphere which may be increased not only by natural but also by anthropogenic emissions. Furthermore, under certain chemical conditions and with a high fraction of organic material, the soil continues to represent a sink for mercury. Changes, particularly of the pH and redox properties, may lead to mobilization, which, however, represents a danger for the ground water and thus transport into aquatic systems rather than a pollution of the vegetation leaf organs. Input into the food chain then occurs via uptake of soil (earthworms) and plant material (Table 8).

5 Conclusion

Studies with respect to the occurrence and behaviour of mercury in the aquatic and terrestrial environment are based on the development of an effective detection system and reproducible analysis. The results have been achieved by applying the cold vapor technique with an accumulation step on gold wool and an AAS system as the detector [32]. It is thus possible to detect Hg concentrations of 0.1 µg/kg in solids and 0.2 ng/l in water with good long-term reproducibility [47]. This is particularly significant for Hg determination in various samples from the Environmental Specimen Bank (ESB). Furthermore, a process has been developed for the determination of methylmercury where after hydrochloric extraction and separation with an ion exchanger, determination as Hg^{2+} is carried out by cold vapor AAS [35]. In this way it is possible to determine the MeHg contents in most matrices of abiotic and biotic systems with a high yield and good reproducibility. Within the framework of international cooperation, a further separation procedure has been developed based on acid distillation, which in combination with the ion exchanger similarly provides reproducible results for the matrices of soil, sediment, soil water and rainwater [36–38].

Transport and accumulation mechanisms and transformation tendencies in marine, limnetic and terrestrial systems were indicated on the basis of the continuous determination of Hg and MeHg load in various samples of abiotic and biotic matrices and from the human food chain at selected sites in the FRG within the framework of the Environmental Specimen Bank and several accompanying studies.

In aquatic systems the sediments are more representative than the water body both with respect to characterization of the pollution as well as concerning possible indications of transformation processes. Transport in the water body is primarily in a particulate form. Corresponding to the different chemism conditions of the sediment, it was possible to observe Hg accumulations and indications of the formation of methylmercury. Apart from the input of methylmercury from atmospheric precipitation and the drainage of terrestrial systems, this toxicological extremely effective compound is taken up into the aquatic food chain and accumulates particularly in fish.

Although no proportional increase of the methylmercury in relation to the total mercury supply was determined in the terrestrial system, nevertheless the significance of the organic substance and the soil reaction with respect to the fixation and transport of mercury and methylmercury within the soil became apparent. Even if formation of methylmercury in the soil cannot be ruled out as a function of certain chemism conditions, the establishment of an equilibrium with the simultaneous input of MeHg from the atmosphere and degradation processes may continuously take place in soil, thus making it difficult to observe this transformation process in the field. The significance of the input of mercury and methylmercury by precipitation became apparent in documenting the pollution of spruce shoots and deciduous leaves.

Complex interactions between the individual compartments of the terrestrial and aquatic system and the atmosphere make it almost impossible to unambiguously follow the path of mercury into the food chain.

Acknowledgement: Mercury, mainly the speciation of mercury and the transformations processes of mercury in the environment was always of special interest to Dr. M. Stoeppler. In this report we have tried to summarize some of the results from the investigations on mercury. We express our sincere thanks for his continuous help in solving scientific problems and supplying resources, without which we could by no means accomplish such a tremendous study.

Most results emerge from investigation of samples from the Environmental Specimen Bank Programm and some accompanying projects. Therefore, the financial support by the Federal Ministry of Environment, Nature Protection and Nuclear Safety (BMU) and by the Federal Environmental Agency of the Federal Republic of Germany (UBA, Berlin) is gratefully acknowledged.

Another part is the result of detailed investigations during the diploma dissertation and the doctoral thesis of the first author, who is deeply indebted to Dr. M. Stoeppler for his continued stimulating interest and discussions in this area.

The authors are also grateful to Dr. M. Horvat, Jozef Stefan Institut, Ljubljana, Slovenia, who participated in the development of analytical methods for mercury speciation within the framework of a bilateral cooperation. Thanks are also due to Dr. M. Lugowska and DI. M. Burow for the analysis of mercury and methylmercury in various materials.

References

1. Greenwood MR (1984) Quecksilber. In: Merian E (ed) Metalle in der Umwelt. Verlag Chemie, Weinheim, p 511
2. Stock A, Cucuel F (1934) Naturwissenschaften 22: 390
3. Winkler HA (1975) Chem Ing Tech 47: 659
4. Tölg G, Lorenz I (1977) Chemie in unserer Zeit 11: 150
5. Wedepohl KH (1969–1978) Handbook of geochemistry, Vol 1–II. Springer, Berlin Heidelberg New York
6. Wedepohl KH (1984) Die Zusammensetzung der oberen Erdkruste und der natürliche Kreislauf ausgewählter Metalle. In: Merian E (ed) Metalle in der Umwelt. Verlag Chemie, Weinheim, p 1
7. Rekolainen S (1985) The deposition of airborne mercury and its effect on mercury sedimentation in Finnish lakes. 5th Int. Conf. Heavy metals in the environment, p 217
8. Drabaek I, Iverfeldt A (1991) Mercury. In: Stoeppler M (ed) Hazardous metals in the environment. Elsevier Science Publishers, Amsterdam (in press)
9. Stumm W, Keller L (1984) Chemische Prozesse in der Umwelt. In: Merian E (ed) Metalle in der Umwelt, Verlag Chemie, Weinheim, p 21
10. Friedrich G, Kulms M (1969) Erzmetall 22: 214
11. Maclatchy JE, Jonasson JR (1974) Geol Surv Cand/Paper p 74
12. Bargagli R, Baldi F (1984) Chemosphere 13: 1059
13. Bargagli R, Iosco F, Barghigiani C (1987) Water Air Soil Poll 36: 219
14. Barhigiani C, Siegel B, Bargagli R, Siegel SM (1989) Water Air Soil Poll 42: 169
15. Kurland L, Faro SN, Siedler H (1960) World Neurol 1: 320
16. Johnels AG, Westermark T (1969) Mercury contamination of the environment in Sweden. In: Miller MW, Berg G (eds) Chemical fallout. Charles Thomas Publishers, Springfield Illiois, p 221
17. Goldwater L (1971) Sci Am 224: 15
18. Clarkson TW (1988) Mercury toxicity. In: Liss AR (ed) Essential and toxic trace elements in human health and disease, p 631
19. Craig PJ (1986) Organometallic compounds in the environment. Longman Group Limited, Essex

20. Jernelöv A, Lann H (1971) Oikos 22: 403
21. Hartung R (1972) The role of food chains in environmental mercury contamination. In: Hartung R, Dinman BD (eds) Environmental mercury contamination. Ann Arbor Sci Publ, Ann Arbor, p 172
22. WHO (1976) Mercury – Environmental health criteria I. Geneva.
23. Paasivirta J, Sarkka I, Surma-Aho K, Humppi T, Kuokkanen T, Marttinen M (1983) Chemosphere 12: 239
24. Fimreite NN (1970) Environ Poll 1: 119
25. Wood JM (1971) Advances Environm Sci Techn 2: 39
26. Johnasson K, Lindqvist O, Timm B (1988) Mercury. National Swedish Environ Protec Board, Report 3524, Solna
27. Hakanson L, Nilsson A, Andersson T (1988) Environ Poll 49: 145
28. Iskandar IK, Seyers JK, Jacobs DR, Keeney DR, Gilmour JT (1972) Analyst 97: 388
29. Revis NW, Osborne TR, Holdsworth G, Hadden C (1990) Environ Contam Toxicol 19: 211
30. Revis NW, Osborne TR, Holdsworth G, Hadden C (1989) Water Air Soil Poll 45: 105
31. Lindqvist O, Johnasson K, Aastrup M, Andersson A, Bringmark L, Hakanson L, Iverfeldt A, Meili M (1991) Mercury in the Swedish Environment – Recent Research on Causes, Consequences and Corrective Methods. In: Water Air Soil Pollution 55: 1–261
32. Stoeppler M (1983) Spectrochim Acta 38: 1559
33. Stoeppler M (1984) Fresenius Z Anal Chem 317: 228
34. Stoeppler M (1990) GIT Fachzeits Lab 7: 872
35. May K, Stoeppler M, Reisinger K (1987) Toxicol Environ Chem 13: 153
36. Horvat M, May K, Stoeppler M, Byrne AR (1988) Appl Organomet Chem 2: 515
37. Padberg S (1991) Doctoral thesis, University of Tübingen, Jüe–2534.
38. Padberg S, Burow M, May K, Stoeppler M (1991) Methylquecksilberbestimmung in Böden. Proceedings 6th CAS, Konstanz April 8–12 1991.
39. Stoeppler M, Schladot JD, Dürbeck HW (1989) GIT Fachzeits Lab 10: 1017
40. Stoeppler M, Schladot JD, Schwuger MJ (1990) Nachr Chem Techn Lab 38: 1228
41. Torres da Silva BP (1989) Doctoral thesis, University of Bonn, Jüe–2265.
42. Ahmed R, May K, Stoeppler M (1987) Sci Tot Environ 60: 249
43. Weber JH, Reisinger K, Stoeppler M (1985) Environ Techn Lett 6: 203
44. Padberg S, Stoeppler M (1988) Tübinger Geogr Stud 100: 1
45. Padberg S, Stoeppler M (1992) Studies of transport and turnover of mercury and methylmercury. Proceedings of Workshop on Toxic Metal Compounds, Interrelation between Chemistry and Biology, Les Diablerets March 4–8 1991 (in press)
46. Welz B (1983) Atom-Absorptions-Spektroskopie, Verlag Chemie, Weinheim
47. May K, Reisinger K, Torres BP, Stoeppler M (1985) Fresenius Z Anal Chem 320: 626
48. May K, Stoeppler M (1984) Fresenius Z Anal Chem 317: 248
49. Ahmed R, May K, Stoeppler M (1987) Fresenius Z Anal Chem 326: 510
50. Westöö G (1966) Acta Scand 20: 2132
51. Millward GE, Bihan A (1978) J Air Poll Contr Ass 35: 1249
52. Dorten W, Valenta P, Nürnberg HW (1984) Fresenius Z Anal Chem 317: 264
53. Ahmed R, Stoeppler M (1986) Analyst 111: 1371
54. Förstner U, Müller G (1974) Schwermetalle in Flüssen und Seen als Ausdruck der Umweltverschmutzung, Springer, Berlin Heidelberg New York
55. Groot AJ (1971) Geol Mijnbouw 50: 393
56. Hellmann H (1970) Dt Gewässerkdl Mitt 146: 160
57. Müller G (1979) Umschau 79: 778
58. Müller G (1983) Bild Wissenschaft 5: 95
59. Hakanson L (1974) Ambio 3: 37
60. Förstner U, Patchineelam SR (1976) Chemikerzeitung 100: 49
61. Förstner U, Wittmann GT (1983) Metal pollution in aquatic environment. Springer, Berlin Heidelberg New York
62. Förstner U (1983) Fresenius Z Anal Chem 316: 604
63. Förstner U, Salomons W (1984) Mobilisierung von Schwermetallen bei Wechelwirkungen mit Sedimenten. In: Merian E (ed) Metalle in der Umwelt. Verlag Chemie, Weinheim, p 171
64. Wood JM, Rosen CG, Kennedey SF (1968) Nature 220: 173
65. Jensen A, Jernelöv A (1969) Nature 223: 753
66. Jernelöv A (1972) Factors in the transformation of mercury tò methylmercury. In: Hartung R, Dinman BD (eds) Environmental mercury contamination. Ann Arbor Sci Publ, Ann Arbor, p 167

67. Spangler WJ (1973) Science 180: 192
68. Schindler JE, Alberts JJ (1975) Verh Int Verein Limnol 19: 2201
69. Reisinger K, Stoeppler M, Nürnberg HW (1983) Fresenius Z Anal Chem 316: 612
70. Fagerström T, Asell B (1973) Ambio 2: 164
71. Wachs B (1982) Münch Beit Abw Fisch Flußbiol 34: 301
72. Padberg S (1987) Diploma dissertation University of Köln
73. Sonderforschungsprojet 327 der Universität Hamburg, Wechselwirkungen zwischen abiotischen und biotischen Prozessen in der Tideelbe
74. Tent L (1983) Vom Wasser 61: 99
75. ARGE Elbe (1990) Schwermetalldaten der Elbe von Schnackenburg bis zu See 1979/80, Hamburg 1980
76. Lugowska M (1988) personal communication
77. Burow M, Stoeppler M (1991) personal communication
78. May M, Stoeppler M (1983) Studies on the biochemical cycle of mercury I. Mercury in sea and inland water and food products. Proceedings Int Conf Heavy Metals in the Environment, Heidelberg, Sept 1983, Vol 1, p 241
79. Brümmer G (1974) Geoderma 12: 207
80. Craig PJ, Moreton PA (1983) Marine Poll Bull 14: 408
81. Craig PJ, Moreton PA (1984) Marine Poll Bull 15: 406
82. Fagerström T, Jernelöv A (1971) Water Res 5: 121
83. Fagerström T, Jernelöv A (1972) Water Res 6: 1193
84. May K (1986) personal communication
85. Stoeppler M, Backhaus M, Burow M, May K, Mohl C (1988) Comparative investigations on race metal levels in brown algae and common (blue) mussels at the same location in the baltic sea and the north sea. In: Wise SS, Zeisler R, Golstein GM (eds) NBS Special Banking 740, Progress in Environmental Specimen Banking, US Department of Commerce, National Bureau of Standards, p 53
86. Stoeppler M (1979) Choice of species, sampling and sample pretreatment for subsequent analysis and banking of marine organisms useful for mercury, lead and cadmium monitoring. In: Luepke NP (ed) Monitoring Environmental Material and Specimen Banking, Martinus Nijhoff Publishers, Boston, p 55
87. Olson I (1984) Bakterien und Pilze – biologische Umwandlung von Metallverbindungen. In: Merian E (ed) Metalle in der Umwelt, Verlag Chemie Weinheim, 10–20
88. Yedilier A, Braun F (1980) Fisch und Umwelt 8: 135
89. Herms U, Brümmer G (1984) Z Pflanzenern Bodenkd 147: 400
90. Brümmer G (1986) Heavy metal species, mobility and availability in soils. In: Bernhard M, Brinckman FE, Sadler PJ (eds) The importance of chemical speciation in environmental processes, Springer Berlin Heidelberg New York, p 169
91. Kerndorff H, Schnitzer M (1980) Geochim Cosmochim
92. Andersson (1979) Mercury in soils. In: Nriagu JW (ed) The biochemistry of mercury in the environment. Elsevier, North Holland, Biomedical Press, Amsterdam New York Oxford, p 79
93. DiGiulio RT, Ryan EA (1987) Water Air Soil Poll 33: 205
94. Nagase H, Ose Y, Saro T, Ishikawa T (1984) Sci Tot Environ 32: 651
95. Woggon H, Klein S, Jehle D, Zydek G (1984) Nahrung 28: 851
96. Rogers RD (1976) Methylation of mercury in terrestrial environment. Proceedings Int Conf Environ Sensing Ass, Meeting Date 1975, Vol 2, p 1
97. Rogers RD (1976) J Environ Qual 5: 454
98. Rogers RD (1977) J Environ Qual 6: 463
99. Wyttenbach A, Bajo S, Tobler L (1986) Aerosols deposits on pine needles. MTAA, 7th Int Conf 1936–1986 Modern Trends on Activation Analysis, Copenhagen, 23.-37.4.1986, p 1097
100. Godbold DL, Hüttermann A (1986) Water Air Soil Poll 31: 509
101. Mosbaek H, Tjell JC, Sevel T (1988) Chemosphere 17: 1227
102. Siegel SM, Siegel BZ (1988) Water Air Soil Poll 40: 443

7 Future Developments

7.1 Standard Reference Materials for the Identification and Determination of Inorganic and Organometallic Compounds of Trace Elements

K.J. Irgolic

The importance of knowledge about the chemical nature and the concentrations of trace element compounds in environmental samples is now widely recognized. Methods are available to identify and determine trace element compounds that are kinetically stable under the conditions imposed by the analytical procedure. The most common techniques for the determination of trace element compounds are discussed. The few standard reference materials with certified concentrations of trace element compounds (DORM-1 dogfish muscel, arsenobetaine; PACS-1 sediment, tributyl tin) are presented.

Introduction

The rapid development of analytical chemistry during the last few decades has led to the production of instruments and methods that allow inorganic analytes present at very low concentrations to be determined even in samples with complex matrices. Concentrations of elements in the microgram per liter or microgram per kilogram range are now routinely determinable. Even lower concentrations pose no insurmountable problems. Trace analysis has become an important part of analytical chemistry and provides data that are used to judge the wholesomeness of food, the purity of drinking water, the extent of pollution of water bodies, soils, and air, the health status of sick and normal people, and the appropriateness of governmental regulations.

The importance of inorganic substances, to which one may count organometallic compounds and complexes of transition elements with organic ligands, to life is aptly expressed by the terms "essential elements" and "toxic elements." Trace elements are important not only for humans, but also for all other living organisms [1]. A new branch of chemistry, known as bioinorganic chemistry, has evolved. Symposia, congresses, and workshops are frequently held to discuss trace element issues, books summarize the knowledge in this area [2–6], and many articles in a variety of journals describe the latest research results. As a logical and necessary extension of the research on trace elements, the inadequacy of total trace element concentrations for an understanding of the interactions between trace elements and organisms is now clearly recognized.

Specimen Banking
Rossbach/Schladot/Ostapczuk (Eds.)
© Springer-Verlag Berlin Heidelberg 1992

Total Trace Elements or Trace Element Compounds?

The spectroscopic methods, in particular atomic absorption and emission spectroscopy, for the determination of inorganic constituents in environmental samples need atoms in the gas phase. Therefore, the analytes are introduced into a high temperature region, in which almost all inorganic and organometallic compounds are decomposed into atoms. Without such thorough destruction of molecular species, these techniques could not provide the information about the concentrations of trace elements in a large variety of samples. Unfortunately, these techniques destroy by design the information that is most important for an evaluation of the impact of an "element" on the biosphere. The "atomic" analytical methods became so commonplace, that even the language was influenced. The terms "essential element" and "toxic element" are creations of the atomic spectroscopic analysts that are widely used and officially accepted in spite of the fact that not elements but compounds of the elements interact with biologically important molecules and maintain through such interactions life processes, impair them, or – in extreme cases – shut them down.

Selenium is one of the newest essential trace elements. Selenium is known to be anticarcinogenic. Selenium has many other biological properties, many of which are still unknown. Selenium is used in medicine [7]. However, in none of these applications or interactions is elemental selenium involved. Selenoamino acids, methylated selenium compounds, selenium-containing lipids, or gluta-thioneperoxidase, an enzyme with four selenium atoms per molecule, perhaps inorganic selenite or selenate, or other yet unknown selenium compounds are the active agents. Obviously, total selenium concentrations are not sufficient to understand the actions of "selenium" on a molecular basis; the chemical nature of the selenium compounds present and their concentrations must be known for this purpose.

Arsenic, an element with a very bad reputation – it is considered to be one of the worst poisons, is ubiquitous and present in high concentrations in marine organisms. Concentrations of 50 mg/kg are encountered rather frequently. Populations with a large portion of their diet derived from marine plants and animals will justifiably worry about the effects of the arsenic consumed with the otherwise tasty seafood. If all the arsenic in the algae, crabs, lobsters, shellfish, and fish were inorganic arsenite containing trivalent arsenic, a seafood diet would certainly not be classified as a health-sustaining diet. Arsenite is the most toxic among the common inorganic arsenic compounds. Fortunately, arsenic is present in these marine organisms in form of simple methylated and more complex organic compounds. Arsenobetaine, trimethyl(carboyxmethyl)arson-ium zwitterion, is in many marine samples the predominant arsenic compound. That arsenobetaine is not toxic even at high doses [8, 9], is an important experimental discovery for all people, who consume seafood by choice or by necessity. The total arsenic concentration in marine samples is a very poor indicator of the safety of seafood in regard to arsenic.

Mercury is another element, the action of which is strongly dependent on its chemical form. Common knowledge suggests, that exposure to mercury vapors should be avoided or kept to the shortest possible time. Whereas elemental mercury is toxic, alkylated mercury compounds, particularly methyl mercury compounds and dimethyl mercury, are much more insidious and much more toxic and destructive than elemental mercury [10–12]. Considerable advantages accrue in judging the safety of food items that are known to contain mercury, when the nature and concentrations of mercury compounds in these items are also defined.

An excellent example for the importance of trace element compounds in human nutrition is a case reported and discussed by Prasad [13]. Iranian villagers were diagnosed of suffering from zinc deficiency in spite of the analytically ascertained fact that food and water contained sufficient zinc to supply the required daily dose. The zinc, however, was in a chemical form that was not resorbed in the intestinal tract. Phytic acid present in the diet combined with the zinc cation: the resulting zinc phytate could not cross from the intestine into the blood stream. The results of the determinations of total zinc in the diet of these Iranian villagers led there to a false security with respect to the adequacy of their zinc intake. Only the identification of the zinc phytate can establish the reason for the observed deficiency symptoms.

Even among purely inorganic compounds drastic differences in biological activities can be found. Chromium as trivalent element and present in food as the "glucose tolerance factor" is an essential element. Chromium as chromate, in which chromium is in hexavalent form, may destroy cells because of its high oxidizing power, may be a strong carcinogen, or may be without effect. The solubility properties of a chromate-containing compound will determine the fate of an affected cell. Soluble chromates, such as alkali metal chromates, will oxidatively destroy a cell. Chromates with low solubility, such as zinc chromate, are carcinogenic, whereas very insoluble chromates, such as barium chromate, appear to be inoccuous [14, 15]. Without information about the type of chromium compound, to which an organism is exposed, little can be said about the effects such an exposure may have.

Similar examples could be cited for many other elements that are essential or toxic. The examples given clearly indicate, that total trace element concentrations cannot be used to predict the impact of exposure to such elements. The importance of knowledge about the chemical nature and the concentrations of particular trace element compounds in environmental samples is now widely recognized; unfortunately, total trace element determinations are still the rule and the methods available to identify and quantify trace element compounds are not widely used. The importance of trace element compounds to the environment is lucidly outlined by the report of the Dahlem Workshop on the Importance of Chemical Speciation in Environmental Processes [16]. Even a special term came into use to describe the analytical activities leading to the identification of trace element compounds, the collection of trace element compounds present in the samples, and the processes that change such compounds in a system. This term "speciation" is a chameleon word and should be expurgated from the scientific literature, in order to return [17] to the useful maxim "only one meaning to each term". Even a cursory reading of papers, in which the term "speciation" is present, will lead to the disturbing conclusion that clarity of communication is not enhanced by the liberal use of this term. A definition for the term "speciation" may come from an IUPAC committee.

Identification and Quantification of Trace Element Compounds

A variety of methods are now at the disposal of the analyst for the identification and quantification of a fair number of trace element compounds. The analytical methods developed for organic compounds had to be compound-specific, because knowledge – even quantitative information – about the presence of C, H, and heteroatoms will only in the case of very simple molecules identify a compound unequivocally. The development of such molecule-based analytical methods for inorganic and organometallic compounds [18] was until recently not an urgent concern of analytical chemists. All the inorganic and organometallic compounds that behave similarly to organic compounds, can of course be identified and quantified by the methods that are useful for organic compounds. Gas chromatography, mass spectroscopy, and GC-MS may serve as examples of such methods and tetraethyl lead, arsine, and tributylalkyl tin as examples for inorganic and organometallic compounds that can be determined by these methods. Unfortunately, very few inorganic and organometallic compounds present in environmental samples possess the volatility required by these "organic" analysis methods.

Fortunately, a number of compounds that are not sufficiently volatile can be converted to volatile derivatives through reactions leading to well defined products. These products can then be identified and quantified in the same way as "naturally" volatile compounds. The reduction of methylarsenic acid to methylarsine and the alkylation of the triethyl lead cation to butyl triethyl lead are examples of such reactions.

However, the large majority of inorganic and organometallic compounds of environmental interest are nonvolatile and cannot be converted to volatile compounds without loss of molecular information. Examples of such compounds are methylcobalamin, trimethylselenonium salts, arsenobetaine, alkyl lead cations, most transition metal complexes, and phosphates. Such compounds are best identified and determined by systems consisting of liquid chromatographs and element-specific detectors. The large variety of high-performance column materials for liquid chromatography provides the opportunity to find appropriate combinations of stationary and mobile phases for the successful separation of anionic, neutral, and cationic analytes. The chromatography can be performed with open columns, with HPLC systems, and with IC systems. Flame emission spectrometers, flame absorption spectrometers, graphite furnace atomic absorption spectrometers, direct-current plasma emission spectrometers, inductively coupled argon plasma emission spectrometers, and inductively coupled argon plasma mass spectrometers have been frequently used as single-element-specific detectors or as multielement-specific detectors. The element-specific detectors simplify the chromatography considerably. Only those compounds must be separated from each other that contain the same element to be used for detection. Compounds with different detectable elements need not be separated by chromatography. For example, a mixture consisting of several selenoamino acids, thioamino acids, and amino acids not containing selenium or sulfur must be chromatographed under conditions that allow all selenoamino acids to be separated from each other and – in the case that the thioamino acids should also be identified – all thioamino acids to be separated from each other. However, selenoamino acids need not be separated from thioamino acids and neither of the chalcogenoamino acids must be separated from the amino acids that do not contain selenium or sulfur. Chromatography element-specific detector systems are described in the literature and have been used to identify and determine arsenic, sulfur, selenium, mercury, tin, silicon, lead, chromium, iron, cobalt, copper, and phosphorus compounds [19–21]. Microwave-induced plasma emission spectrometers are more useful as element-specific detectors in gas chromatography than in liquid chromatography. Generally, the microwave-induced plasmas cannot be sustained when exposed to volumes of liquids corresponding to flow rates commonly employed in liquid chromatography.

The detection limits for these combined methods – often called hyphenated techniques – are not as low as the detection limits obtainable with the spectrometers alone. The chromatography dilutes the analyte and thus more of the analyte is required to be detectable. Although these hyphenated techniques have been used for the determination of a variety of trace element compounds, the techniques do not enjoy universal applicability. A sample, in which trace element compounds should be determined, is often extracted with water or with an aqueous/organic solvent to concentrate the analytes and perhaps separate them from part of the matrix. In many cases, the extract must be cleaned up, for instance, by extractive procedures or ion chromatography, before the analytes are separated by liquid chromatography [22]. All these processes change the environment, in which the trace element compounds reside. These changes must not change the composition of the trace element compounds and must not change their redox state. In order to identify and determine trace element compounds with these methods, the compounds must be kinetically stable under the experimental conditions. This requirement is met by many compounds of elements that have their place in the periodic table in the border region between metals and nonmetals. Most of the transition metal compounds and complexes of environmental interest do not fulfill this requirement and cannot be identified by these methods. In-situ methods should be invented for the kinetically labile compounds.

Standards Reference Materials for the Identification and Determination of Trace Element Compounds

Although the determinations of trace element compounds is a rather mature art and perhaps a mature science, all is not well even with these analytical tasks. Round-robbin experiments often produce very divergent results. The analytical error of a determination of a trace element even in a not so complex matrix increases exponentially with decreasing concentration of the trace element. The result may be orders of magnitudes from the actual value [23]. To avoid such discrepancies as much as possible, standard references materials are offered by organizations such as the US National Institute of Standards and Technology [24]. The concentrations of trace elements in these standards were determined by a series of independent methods and are reported as reference values with standard deviations. An analytical procedure can now be tested with these standards to check on the reliability of the procedure. However, total certainty about the accuracy of an analytical result can only be obtained, when the matrices of the sample and the standard are matched. The identification and determination of a trace element compound is a more complex task than the determination of the total concentration of a trace element. Reference materials for this more complex task are much needed. Unfortunately, very few reference materials exist, in which trace element compounds were identified.

Arsenobetaine was identified in the dogfish muscle reference material DORM-1 (National Research Council of Canada) by high-performance liquid chromatography with arsenic-specific detection by inductively coupled argon plasma-mass spectrometry, thin-layer chromatography, and electron-impact mass spectrometry [25]. Arsenobetaine was quantified in this material by HPLC-ICP/MS and graphite furnace atomic absorption spectrometry. Arseno-betaine was the major arsenic compound in the dogfish muscle and accounted at a concentration of 15.7 ± 0.8 microgram arsenic per gram of sample for 84% of the total arsenic.

The tributyl tin concentration in the sediment reference material PACS-1 (National Research Council of Canada) was determined by extraction/gas chromatography of tributyl tin chloride (1.08 ± 0.31 microgram tin per gram sediment) [26], by extraction/ionspray mass spectrometry/mass spectrometry (1.29 ± 0.07 microgram/g) [27], and by HPLC/ICP-MS (1.18 ± 0.15 microgram/g) [28]. Dibutyl tin was also quantified in this sediment (1.13 ± 0.30 microgram/g by extraction/GC; 1.19 ± 0.14 microgram/g by HPLC/ICP-MS).

The NIST standard reference material "Oyster Tissue" with a certified total arsenic concentration of 13.4 ± 1.9 mg/kg was also investigated with respect to arsenic compounds [29]. Boiling water extracted almost all of the arsenic from the sample. After mineralization of the extract an arsenic concentration of

13 mg/kg was found by hydride generation [30]. The not mineralized extract did not contain simple methylated arsenic compounds, but produced a signal for arsenate (0.4 mg As/kg). Treatment of the oyster tissue with a hot 4 molar sodium hydroxide solution should convert arsenobetaine into trimethylarsine oxide and ribosyldimethylarsine oxides into dimethylarsinic acid [31]. These arsenic compounds should be reduced in the hydride generation system to trimethylarsine and dimethylarsine, respectively. When the acidified extract was reduced, only a signal for arsenate was obtained (2.6 mg As/kg). The observations that all the arsenic in the oyster tissue is extractable by hot water, but no signs of the presence of simple methylated arsenic compounds or arsenobetaine could be found, is puzzling.

Trace Element Compounds and Specimen Banks

The standard reference materials with certified concentrations for total trace elements are now used by analysts all over the world to evaluate their analytical procedures and increase the reliability of their results. The identification and quantification of trace element compounds in environmental samples, a much more complex undertaking than the determination of total trace element concentrations, would be greatly aided by standard reference materials certified for trace element compounds. The first step in the development of such standards is the identification of trace element compounds in suitable reference materials certified for total trace elements. This work has already begun and should continue at an accelerated pace. Samples stored in specimen banks are ideal materials for such investigations. The methods for the identification and quantification of trace element compounds that are now available will certainly be refined and new and better methods will be developed. The historical trend of total trace element concentrations, for instance in human sera, clearly indicates [23], that better methods give lower concentrations for many elements. Such method-dependent changes can be easily recognized, when the same sample remains available over extended periods of time. Specimen banks will greatly facilitate the evaluation of the stability of trace element compounds in reference materials. A standard reference material must not only have identified trace element compounds at well defined concentrations at the time the analyses were performed, but also at all later times, at which such a material is used as a standard. Therefore, the stability of each certified trace element compound in a reference material must be checked over long periods of time. No other samples are better suited for such procedures than samples stored and protected in a specimen bank.

Trace element compounds will certainly increase in importance in the future. The cycling of elements in the environment, the detrimental and beneficial influences of trace elements on biological systems, therapeutic interventions based on trace elements, and many other aspects of the biochemical interactions of trace elements with biomolecules important for life cannot be understood without detailed knowledge about trace element compounds and their transformations in biological and abiological systems. Specimen banks have a vital role in these important developments concerning trace element compounds.

References

1. Schröder HA (1978) The trace elements and man. The Devin Adair Company, Old Greenwich, Connecticut, USA
2. Irgolic KJ, Martell AE (Eds) (1985) Environmental inorganic chemistry. VCH Publishers, Deerfield Beach, Florida, USA
3. US National Academy of Sciences (1972–1978) Series on "Medical and biological effects of environmental pollutants", Volumes on chromium, copper, lead, manganese, nickel, selenium, vanadium, arsenic, mercury, platinum-group metals; Washington, DC, USA
4. Craig PJ (Ed) (1986) Organometallic compounds in the environment: principles and reactions. John Wiley, New York
5. Xavier AV (Ed) (1986) Frontiers in bioinorganic chemistry. VCH, Weinheim, Germany
6. Merian E (Ed) (1984) Metalle in der Umwelt: Verteilung, Analytik und biologische Relevanz. Verlag Chemie, Weinheim, Deutschland
7. Wendel A (Ed) (1989) Selenium in biology and medicine. Springer, Berlin Heidelberg New York
8. Vahter M, Marafante E, Dencker L (1983) Sci Total Environ 30: 197
9. Irvin RT, Irgolic KJ (1988) Appl Organomet Chem 2: 509
10. D'Itri (Ed) 1972) The environmental mercury problem. CRC Press Cleveland, Ohio, USA
11. Friberg L, Vostal J (1972) Mercury in the environment. CRC Press Cleveland, Ohio, USA
12. Rabenstein DL (1978) J Chem Ed 55: 292
13. Prasad A (Ed) (1976) Trace elements in human health and disease; Vol 1: Zinc and copper. Academic, New York
14. Bianchi V, Levis AG (1985) Metals as genotoxic agents: the model of chromium; in Ref. 2: 447
15. Patty FA (Ed) (1963) Industrial Hygiene and toxicology. Interscience Publishers, New York, p. 1021
16. Bernhard M, Brinckman FE, Sadler PJ (Eds) (1986) The importance of chemical "speciation" in environmental processes. Springer Verlag, Berlin
17. Irgolic KJ (1989) Chem Speciation Bioavail 1: 127
18. Irgolic KJ (1989) Chem Speciation Bioavail 1: 47
19. Irgolic KJ, Brinckman FE (1986) Liquid chromatography element-specific detection systems for analysis of molecular species; in Ref. 16: 667
20. Irgolic KJ (1991) The determination of arsenic and arsenic compounds. In: Stöppler M (ed) Hazardous metals in the environment, Elsevier, Amsterdam (in press)
21. Krull IS (Ed) (1992) Trace metal analysis and speciation. Elsevier, Amsterdam
22. Francesconi K, Micks P, Stockton RA, Irgolic KJ (1985) Chemosphere 14: 1443
23. Tölg G (1989) Kontakte (Merck) Issue 2: 21
24. NIST, Standard reference materials catalog 1990–1991, p. 50
25. Beauchemin D, Bednas ME, Berman SS, McLaren JW, Siu KWM, Sturgeon RE (1988) Anal Chem 60: 2209
26. Siu KWM, Maxwell PS, Berman SS (1989) J Chromatogr 475: 373
27. Siu KWM, Gardner GJ, Berman SS (1989) Anal Chem 61: 2320
28. McLaren JW, Siu KWM, Lam JW, Willie SN, Maxwell PS, Palepu A, Koether M, Bermann SS (1990) Fresenius J Anal Chem 337: 721
29. Kaise T, Puri BK, Irgolic KJ (1990) unpublished
30. Clark PJ, Zingaro RA, Irgolic KJ, McGinley AN (1980) Intern J Environ Anal Chem 7: 295
31. Kaise T, Yamauchi H, Hirayama T, Fukui S (1988) Appl Organomet Chem 2: 339

7.2 Prompt Gamma Activation Analysis with Cold Neutrons for the Characterisation of Specimen Bank Materials

M. Rossbach

A Prompt Gamma Cold Neutron Activation Analysis (PGCNAA) facility at the ELLA-laboratory of the Research Center, KFA-Jülich, Germany is described emphasizing the experimental progress in conjunction with the use of cold (= slow) neutrons. This "in-beam" technique is nondestructive, has multielement capabilities, and is particularly suitable for analyzing low-Z (constituting) elements in biological and environmental materials. Results are presented for selected certified reference materials and a number of Environmental Specimen Bank (ESB) materials including a candidate reference material "spruce shoots" RMF I and RMF II.

Introduction

The German Environmental Specimen Bank (ESB) project is described in detail by several authors in this volume [1–3] and in [4, 5]. The analytical capabilities for the characterisation of the stored materials of the ESB with respect to inorganic constituents include advanced physico-chemical techniques such as atomic absorption and -emission, voltammetry, mass-spectrometry, or neutron activation analysis with oligo- or multielement capabilities. The routine application of these techniques generally leads to a highly reliable determination of toxic elements such as Hg, Cd, Pb, Ni, and As and a number of essential elements, e.g. Co, Zn, Cu, Mn, and Cr [6]. Other important minor and trace elements such as Ca, K, Mg, V, Se, Mo, Sn etc. are only determined sporadically or for lack of appropriate methodology, not at all.

A strong stimulating aspect arising from ongoing specimen banking is the development and improvement of analytical techniques capable of analyzing more and hitherto unrecognized elements in ESB materials, hence, increasing our knowledge of the chemical composition and the behavior of elements and compounds in the environment. The protocollized sampling and strictly controlled storage of specimens particularly selected for its strong environmental evidence, liberates the analyst from taking undue care over these steps and provides full confidence in the quality of the unchanged sample material.

The need for additional information on bulk elements in ESB materials initiated an attempt to install a device for Prompt Gamma Analysis of samples activated with cold neutrons (PGNAA), at the ELLA external neutron laboratory of the FRJ-2 reactor, KFA-Jülich [7]. This nuclear analytical technique has

Specimen Banking
Rossbach/Schladot/Ostapczuk (Eds.)
© Springer-Verlag Berlin Heidelberg 1992

been known for almost thirty years [8, 9] to be particularly suitable for the quantification of low-Z elements such as H, B, C, N, S etc. and some trace elements (Cd, Gd) in geological and biological materials. Unfortunately the scarcity of filtered neutron beams in the past did not allow a rapid development of the technique until recently. The installation of cold sources and ^{58}Ni-coated beam tubes at several research reactors in the world led to enhanced development of the method. Multielement applications are described for geological [e.g. 10–12], industrial [13], fuel [14], and biological materials [15–18]. Additionally interesting applications are single element determinations in special materials such as the in vivo determination of N, Ca, K, or Cd in patients [19], sulfur in coal [20], or boron in atmospheric particulates and gas-phase species [21].

The method is based upon the registration of spontaneously emitted charac-teristic gamma rays of isotopes induced by the absorption of neutrons impinging the sample. It is an "in beam" spectroscopic method which is based on nuclear properties alone and therefore independent of the chemical form of the elements. It is nondestructive, which means that no chemical treatment of the sample is necessary and thus allows the same aliquot of a sample to be analyzed by the combination of several methods [22].

Specific problems and improvements of the method with particular emphasis on the use of cold neutrons are addressed in a recent publication [23]. The development of a low background producing irradiation chamber lined with a ^{6}LiF close sample/detector geometry and a technique for the preparation of flat (≤ 1 mm thickness), large diameter (> 30 cm^2) biological samples improved the sensitivity of several elements considerably. An additional increase in sensitivity will be obtained by the foreseen implementation of a Compton suppressor made from BGO (Bismuth-germanate) into the counting system.

Materials and Methods

The details of the instrument, the counting conditions and the preparation of samples and standards are described in [23]. To demonstrate the multielement capabilities of the method several Certified and Standard Reference Materials have been analyzed (SRM 1572, SRM 1575, SRM 1549, RM 8431a, SRM 1577a, and NIES No. 8).

From the specimen bank materials, two spruce shoot specimen RMF I and RMF II used as internal reference materials are characterized and additional data on brown algae, poplar leaves, beech leaves, and bream are presented. The biological materials were dried as received (fresh, homogenized, and in frozen state) in a desiccator for several days over H_2SO_4 until constant weight was indicated. Some volatile organic compounds such as ethers, alcohols, or esters must have been trapped in the H_2SO_4 turning the top layers of the liquid into a dark brown fluid: results from the oxidation of organic material. Water loss was monitored and used to relate concentrations to the fresh weight.

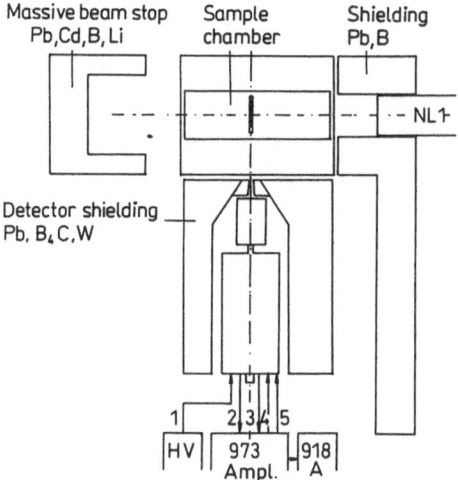

Fig. 1. Schematic setup of the PGCNAA device at ELLA, KFA Jülich

The samples were pressed into flat slabs of ca. $1 \times 47 \times 64$ mm with the help of a heatable pill press, which contained 100 mm diameter dyes ($\leq 50\,°C$, 15 tonnes). The sample slabs were then weighed and put into clean Teflon bags (CRP Inc. Ronkonkoma, N.Y.) and heat sealed. A specially designed irradiation chamber was used to position and expose the samples to the uniform neutron field of 9×10^9 N/s at a wavelength of 5.1 Å. A high purity Ge-detector (EG & G Ortec) of 20% relative efficiency and 1.9 keV (^{60}Co) resolution was used to monitor the resulting gamma radiation (see Fig. 1). Digitalized spectra were recorded on a Toshiba T 5200 portable PC using Ortec software MAESTRO. The calculation of concentrations in the sample materials was carried out manually using synthetic standards prepared from mixtures of high purity metals and compounds with spectroscopic grade cellulose powder of the same size and shape as the samples. In some instances concentrations from CRM's were used for calculation. 237.5 mg of Ti (SRM 354) was used as a flux monitor to detect alterations in neutron beam intensity due to changing reactor power and/or influences by other experiments upstream of our beam.

Results and Discussions

During one fuel cycle of 3 weeks the variation of the neutron flux determined by repeated measurements of the Ti flux monitor remained stable to within 3.9–8.2%.

Background count rates were determined while irradiating a blank Teflon bag. One of the major problems in PGCNAA is the scattering of neutrons in the sample (particularly at hydrogen). The scattered neutrons produce gamma rays by reaction with sample supporting and surrounding materials and hence, the background determination does not really represent substractable background

Table 1. Results of several background measurements under various conditions (counts/s)

Element [KeV]	B [478]	Annih.Rad [511]	C [1261]	F [1633]	Al [1779]	N [1884]	H [2223]
No. 235 600 s no bag	5.38	62.9	1.94	0.63	1.61	4.81	1.91
No. 237 old bag 600 s	17.9	64.07	2.86	8.17	2.69	4.82	2.39
No. 236 new bag 600 s	18.3	63.99	2.89	7.73	2.32	5.19	2.64
No. 248 new bag 900 s	12.97	51.28	2.90	7.69	2.9	4.92	2.23
No. 251 new bag 900 s	9.44	50.34	3.04	7.76	2.75	5.65	2.18

No. 254 new bag 900 s	11.08	51.44	2.87	7.71	2.84	5.11	2.17
No. 257 new bag 900 s	9.31	51.04	2.74	8.08	2.75	4.95	2.14
No. 268 new bag 900 s	10.38	50.99	3.05	7.75	3.07	5.53	2.07
No. 262 new bag 60 500 s	9.58	49.12	2.71	7.81	2.92	5.13	1.88
MW ± S excluding Nr. 235	9.96 ± 7.6%	51.0 ± 2.2%	2.88 ± 4.2%	7.84 ± 2.3%	2.78 ± 7.96%	5.12 ± 5.8%	2.21 ± 10%
=	1.96 µg		5.47 mg	33.3 µg	64.1 µg	5.86%	0.48 mg
for comparison values given by NIST 1982	2.73 µg		5.1 mg		511 µg	3.48%	2.54 mg

for a sample measuring process. Although the effect is generally small for very thin samples it's magnitude can vary from material to material. Results for various background measurements are listed in Table 1.

In Table 2 results from six reference materials are compiled along with given values as far as they were available. The standard deviation corresponds to 1Σ of two or more independent measurements. As the time for the irradiation per sample is 20 000–60 000 s the number of investigated aliquots per specimen is rather limited.

For reference values in the investigated materials the following sources have been used: Certificate of analysis as issued by the supplier [24–29], compilation of trace elements in different SRM's by E.S. Gladney et al. [30, 31], and a recent compilation issued by the IAEA, Vienna [32].

From these compilations it is clear that for C, H, N, F, B, and V no certified values exist for the six investigated materials. The values in parenthesis are from Gladney's tabulation or information values only. Therefore, C, H, and N results are not readily comparable. Sulfur values agree within the error bars for the three certified materials. Phosphorus is too high in pine needles and too low in milk powder; the reason could be the rather poor peak statistics in the vicinity of the strong hydrogen line at 2223. keV. Fluorine data are still doubtful because the large background contribution from ^{6}LiF of the irradiation chamber had to be substracted from the signal; the challenge of flourine determination in biological materials still requires further research. Chlorine is very sensitively detected by PGCNAA and therefore contamination problems are severe with samples, standards, and particularly with the Teflon bags supporting both. Agreement with certified values is quite satisfying, however.

Two different values for boron in SRM 1575 are cited in [30]. Our value of 15.1 ppm tends to be the higher of the two, whereas a reference value from [18] for RM 8431a is 40% higher than ours. Boron in ESB-materials will be of particular interest in future investigations because recent result suggest that B is not only essential for plants but is of particular nutritional value to humans under certain metabolic stressful situations [33]. All six potassium values agree well with the certified concentrations. Vanadium with the rather weak gamma line at 646 keV is suspect of interference from other elements such as K, or Cu with higher abundance but lower capture cross section. Their influence and possible corrections will be further investigated. The detection sensitivity of the element in the sub-ppm level, however, remains and will be used for more reliable analysis in the future. Other elements such as Gd, W, or Ti have been detected but due to the absence appropriate standards could not be quantified yet. Samarium was evaluated on the basis of a non-certified value (0.052 μg/g) in Citrus leaves (SRM 1572). Using this approach the noncertified value in NIES No. 8 could not be verified.

Cadmium is very sensitively detected by PGCNAA. However, the concentrations found in the reference materials do not correspond very well to the certified values. Particularly in the materials with very low Cd concentrations a contribution from contamination cannot be excluded.

Table 2. PGCNAA results from 6 different reference materials in mg/kg. Certified or information values in parenthesis (Ref. 22–30). Values in [ ] from Ref. 18

Element [keV]	Citrus 1. SRM 1572	Pine n. SRM 1575	Vehicle exh. NIES No. 8	Milk pow. SRM 1549	Bovine 1. SRM 1577a	Mixed Diet RM 8431a
C [%] 1261.	48.8 ± 7.4%	44.9 ± 3.9% (51.6)	21.6 ± 5.1%	39.6 ± 5.1%	55.5 ± 5.05%	45.2 ± 8.2% [47.9]
H [%] 2223.	5.84 ± 7.0%	6.03 ± 5.3% (6.4)	2.91 ± 7.1%	5.85 ± 6.1%	6.77 ± 10%	6.58 ± 7.2% [6.67]
N [%] 3577.	0.38 ± 14.2% (2.86)	1.03 ± 12% (1.2)	0.7 ± 5.8%	0.317 ± 9.4%	8.64 ± 6.4% (10.7)	2.93 ± 13.4% [3.28]
S [%] 841, 2379	0.44 ± 19.3% (0.407)	0.112 ± 3.2% (0.104)	1.63 ± 17.5%	0.314 ± 18.4% (0.351)	0.818 ± 14.5% (0.78)	0.247 ± 18.3% [0.2130]
K [%] 7770, 1158	1.84 ± 13% (1.82)	0.378 ± 16.1% (0.37)	0.121 ± 11.8% (0.115)	1.52 ± 9.2% (1.69)	0.97 ± 8,0% (0.996)	0.7326 ± 3.0% (0.79)
Cl [%] 1164, 1950	0.0513 ± 14.8% (0.0414)	0.0363 ± 13.0% (0.04)	0.0978 ± 8.2%	0.9311 ± 7.4% (1.09)	0.2581 ± 6.1% (0.28)	0.4878 ± 5.3% [0.49980]
P 2154.	n.d.	1433 ± 33% (1200.)	1595 ± 38.5% (510.)	5500 ± 30.1% (10600.)	n.d.	1500 ± 23.1% (3320.)
F 665, 1633	1.63 ± 2.5%	ref. value (3.7)	2.88 ± 27.8%	0.79 ± 11.7% (0.2)	12.0 ± 50%	5.73 ± 23.1%
B 478.	33.4 ± 4%	15.06 ± 4.3% (1.7, 9.0)	1.38 ± 6.4%	0.0746 ± 7.8%	3.27 ± 2.24%	2.31 ± 4.5% [3.87]
V 646.	0.64 ± 15.6% (0.4)	0.48 ± 18%	n.d.	0.182 ± 12.1%	0.517 ± 16.8%	0.37 ± 23.7%
Sm 738.	ref. value (0.052)	0.0078 ± 13.1%	0.075 ± 12.8% (0.2)	0.0017 ± 34.6%	0.024 ± 15%	0.021 ± 15.2%
Cd 558.	0.021 ± 8.3% (0.03)	0.045 ± 4.5% (<0.5)	1.2 ± 7.5% (1.1)	0.00177 ± 31% (0.0005)	0.573 ± 13.8% (0.44)	0.072 ± 24.8% (0.042)

To conclude it can be stated that the attempt to use this relatively un-explored technique in conjunction with cold neutrons for multielement deter-mination in biological materials was successful and it can be applied to samples from the ESB.

For the two candidate reference materials RMF I and RMF II, spruce shoots were taken in 50 kg amounts from two sites of the regular specimen bank collecting areas "Berchtesgaden" and "Warndt, Saarland". The material was sampled according to an existing sampling protocol [34].

Two years old twigs were cut from selected trees of *picea abies* (norwegian spruce) ca. 8 m above ground. Particular care was taken not to contaminate the material during the sampling process. The shoots were collected from the branches by cutting with a titanium knife and immersed immediately in ca. 5 kg portions into liquid nitrogen. Transport from the sampling site to the processing and storage laboratory, the ESB in Jülich, was carried out under liquid nitrogen vapor conditions ($\sim$ 140 K).

At the ESB the material was homogenized using the cryogenic mill "Cryo Palla" that is described in [35]. The material was then packed in 200 g portions in clean Duran glass (Schott) bottles and stored at 140 K until further use. Homogeneity studies using solid sampling AAS proved that both materials met reference material criteria for many elements. Particle size distribution was investigated for elements and proved to be excellent. This material was dried in a desiccator over H_2SO_4 for almost a week until constant weight was obtained. The water loss factor and the residue after ignition can be seen from Table 3. The dried materials were pressed into slabs of 700.0 mg, 1053.0 mg, and 1284.0 mg for RM I and 680.0 mg, 1208.0 mg, and 1625.9 mg for RMF II all $\leq$ 1 mm thickness as described above.

Several measurements of these aliquots were performed in the PGCNAA facility at the ELLA laboratory for the period 20 000–60 000 s. By standard comparison and background correction the results were calculated and are listed in Table 4. Two sources for comparison of the data are cited: NIST, Gaithersburg, that performed an INAA characterization of the two materials in a collaborative effort and values from our own institute determined by GF-AAS and voltammetry. Unfortunately only a few elements from PGCNAA analysis are covered by other techniques, therefore no comparison data can be obtained for elements such as C, H, N, S, Si, P, B etc.

Potassium values tend to correspond well with the given data and agree-ment is also excellent with values from our own laboratory (0.55 $\pm$ 2.7% and

Table 3. Characteristic properties of the two candidate ESB-reference mater-ials RMF I and RMF II

Material	Water loss factor	Residue after ignition
RMF I	1.68	17.03 mg/g fresh
RMF II	1.73	31.23 mg/g fresh

Table 4. Results from two candidate reference materials, RMF I and RMF II (spruce shoots) from the Environmental Specimen Bank, Germany in mg/kg dry weight

Element	RMF I MW ± % this work		MW ± % other sources		RMF II MW ± % this work		MW ± % other sources	
C:	61.5%	4.9			56.5%	3.25		
H:	6.5%	5.4			6.44%	11.2		
N:	2.2%	14.4			2.1%	23.0		
K:	0.54%	8.0	0.502ª	3.39	0.56%	18.8	0.556ª	6.5
Ca:	0.54%	10.5	0.536ᵇ	1.7	0.523%	7.3	0.582ª	10.8
Mg:	888.0	7.0	1440.0ᵇ	0.7	0.1576%	30.0	880.0ᵇ	2.3
Al:	904.0	5.4	54.1ª	2.6	925.7	11.9	344.4ª	1.0
S:	853.0	9.3			1943.0	28.0		
Si:	750.0	4.0			2190.0	9.9		
Cl:	685.0	20.0	556.0ª	1.0	1068.0	29.7	844.0ª	13.0
P:	420.0	22.0			659.0	26.0		
B:	10.8	27.0			11.8	0.5		
Ti:	8.44				25.0	18.0		
F:	5.82	14.0			5.5	18.0		
J:	2.1	3.0			1.76	6.5		
Gd:	0.875	11.0	0.165ª	1.5	1.16	18.3		
V:	0.161	36.0			0.149	26.8	1.41ª	21.9
Se:	0.029	11.7	0.086ª	3.0	0.043	3.6	0.096ª	22.9
Cd:	0.0185	13.0	0.02ᵇ	15.0	0.324	7.3	0.37ᵇ	8.1

ª INAA values from NIST, Gaithersburg, Dr. R. Zeisler;
ᵇ AAS values from ICH 4, KFA Jülich, Fr. DI C. Mohl

Table 5. PGCNAA values in four ESB materials in mg/kg^a dry weight.

Element	algae III/6, [*fucus vesiculosus*] MW ± S	82 bream I, [*abramis brama*] MW ± S	85 poplar 1. I, 87 [*populus nigra*] MW ± S	beech 1. 87 [*fagus sylvaticus*] MW ± S
H [%]	4.37 ± 0.3	6.17 ± 0.23	5.39 ± 0.39	6.37 ± 0.3
C [%]	34.9 ± 2.6	34.1 ± 1.3	42.4 ± 2.2	43.7 ± 3.5
N [%]	7.31 ± 1.0	9.34 ± 1.1	1.26 ± 0.07	1.7 ± 0.06
S [%]	2.18 ± 0.37	0.77 ± 0.03	0.422 ± 0.06	0.199 ± 0.02
K [%]	2.57 ± 0.21	1.02 ± 0.03	1.66 ± 0.21	0.81 ± 0.09
Ca [%]	5.3 ± 0.69	7.12 ± 1.5	2.75 ± 0.22	1.03 ± 0.08
Al	17.6 ± 0.4	27.2 ± 0.43	500. ± 15.8	685. ± 53.5
Cl	29300. ± 3100	5540. ± 589.	2800. ± 329.	2425. ± 77.8
B	101.7 ± 5.1	0.43 ± 0.14	66.7 ± 4.7	1.33 ± 0.3
Ti	12.4 ± 2.35	6.55 ± 1.7	3.45 ± 0.83	4.78 ± 0.91
F	(23.4)	(0.52)	(2.0)	4.52 ± 0.57
Li	(0.392)	(0.44)		(2.1)
Sn	(0.648)	(0.119)	(0.0497)	(0.037)
V	1.37 ± 0.12	0.627 ± 0.08	1.02 ± 0.046	n.d.
Sm	0.09 ± 0.003	0.028 ± 0.0087	(0.0874)	(0.019)
Cd	0.581 ± 0.048	(0.67)	1.27 ± 0.02	0.117 ± 0.016

^{a} Values in parentheses are the result of only one determination

Table 6. Values for comparison in *fucus vesiculosus*

Element	Value	
B	164.[a]	120.[b]
H%	4.2[a]	4.1[b]
Cl%	2.8[a]	0.47[b]

[a] From ref. 36
[b] From ref. 37

0.54 ± 3.7, respectively). Calcium values agree quite well compared to values from IPC-4: 0.51 ± 3.9% and 0.49 ± 2%, respectively. With Mg the situation seems to be obscured. Agreement would be good if values for the two materials could be exchanged. No errors could be detected in the calculated PGCNAA values. The materials must therefore be reanalyzed with regard to Mg particularly.

Aluminium values as with the reference materials from Table 2 are too high showing a false background correction; this is due to the influence of scattered neutrons in the irradiation chamber made of ^{6}LiF lined aluminium. These values have to be reconsidered. Chlorine values, although somewhat high, are still within the error bars given. Vanadium agrees well in RMF I but seems to be a factor of 10 too low in RMF II. No reason for this apparent discrepancy could be detected. For selenium, both values are more than a factor 2 lower than the NIST values. Se was evaluated using a rather low energy line at 238.9 keV. As PGCNAA spectra are pretty complex in this energy region interferences from other elements (e.g. Ag at 236.7 keV or Yb at 241.8 keV) cannot be totally excluded. Future consideration will be devoted to such effects and their possible correction is aspired. Cadmium is a very sensitive element and agrees very well with the previously determined data at our own institute.

In Table 5 values for four additional ESB materials brown algae [*Fucus vesiculosus*] (III/6 Dec. 82, Nordstrand), bream I [*Abramis brama*] (Bodensee, Langenargen 1985), poplar leaves I [*Populus nigra*] (Saarland Hostenbach 1987) and beech leaves [*Fagus sylvaticus*] (Bornhöved 1987) are reported. No data for comparison can be given as no published source could be identified for those materials except for three elements in *Fucus vesiculosus* from 1968 (see Table 6).

Conclusion

The advantage of using cold neutrons in prompt gamma neutron activation analysis as augmented sensitivity and low background conditions have been profitably used to demonstrate the multielement capabilities of PGCNAA in application to Environmental Specimen Bank materials.

The implementation of the main constituting elements such as H, C, N, S as well as several hitherto unrecognized elements such as Si, B, F, or Li considerably enlarges the element spectrum for which ESB materials can be characterized and thus provides the ecologist with new and informative data for all kinds of applications. The nitrogen data e.g. can be used to deduce protein content in the fresh materials by applying a conversion factor of $6.25 \times N$ and the water loss factor as determined while drying. The fate and behavior of elements and compounds in ecosystems needs to be monitored for, e.g. risk assessment and mass balance studies, accumulation or depletion of elements equally important for agriculture and in population studies. The analytical capabilities associated with the German Environmental Specimen Bank are therefore continuously improved to provide a reliable basis for legislation and scientific investigations related to environmental quality.

Acknowledgement: Dr. M. Stoeppler, group leader of the section "trace metals in the environment" and strong promotor of the specimen bank activities at the Institute of Applied Physical Chemistry of the KFA-Jülich, has not only supervised my thesis in the years 1980–1985 but strongly promoted my scientific ambitions in various ways. I am very grateful for his mutual interest, his continuing ability for scientific discussion, and his many helpful hints. Additionally I would like to thank Dr. R. Zeisler, head of the Chemistry unit of the Agencies Laboratories at Seibersdorf, IAEA, Austria, and former research scientist of the nuclear analytical group at NIST, Gaithersburg, USA, for his stimulation and many fruitful collaborations in PGCNAA during the last 4 years and for the INAA values in RMF I and II.

References

1. Boehringer U (1991) Specimen Banking: Environmental Monitoring and Modern Analytical Approaches. Springer, Berlin Heidelberg New York
2. Kemper FH (see Ref 1)
3. Schladot JD (see Ref 1)
4. Stoeppler M, Schladot J.-D, Dürbeck HW (1989) GIT Fachz. Lab. 10: 1017
5. Stoeppler M, Schladot J.-D, Dürbeck HW (1989) GIT Fachz. Lab. 11: 1119
6. Stoeppler M (1989) In: Said HM, Rahman MA, D'Silva LA (eds.) Elements in Health and Disease 2nd Int. Conf. on Elements in Health and Disease Feb. 87: 289
7. Alefeld B, Duppich J, Schärpf O, Schirmer A, Springer T, Werner K (1988) SPIE, Vol. 183, Thin Neutron Optical Devices 75
8. Lussie WG, Brownlee JL Jr., (1965) Proc. Conf. Modern Trends in Activ. Anal. 194
9. Isenhour TL, Morrison GH (1966) Anal. Chem. 38: 162
10. Graham CC, Glascock MD, Carni JJ, Vogt JR, Spalding TG (1982) Anal. Chem. 54: 1623
11. Glascock MD, Coveney RM Jr., Tittle CW, Gartner ML, Murphey RD (1985) Nucl. Inst. Meth. B10/11: 1042
12. Mikesell JL, Senftle FE, Anderson RN, Greenberg M (1989) Nucl. Geophys. Vol. 3, No 4 501–519. Int. J. Radioat. Appl. Instrum. Part E
13. Herzog W (1989) Nucl. Geophys. Vol. 3, No. 4 467–473. Int. J. Radiat. Appl. Instrum. Part E
14. Maly J, Bozorgmanesh H (1984) Report DOE/LC/10900-1572, Order No. DE 84009288 168 pp. Avail. NTIS
15. Failey MP, Anderson DL, Zoller WH, Gordon GE, Lindstrom RM (1979) Anal. Chem. Vol. 51, No. 13 2209
16. Anderson DL, Failey MP, Zoller WH, Gordon GE (1978) In: Chrien RE, Kane WR (eds.) Neutron Capture Gamma Ray Spectroscopy 546–548. Plenum, New York, N.Y.

17. Anderson DL (1983) NBS-Spec. Pub. (U.S.) 656: 123
18. Anderson DL, Cunningham WC, Mackey EA (1990) In: Zeisler R, Guinn VP (eds.) Nucl. Anal. Meth. in the Life Sci. Humana Press, Clifton N.J. 613
19. Chettle DR, Fremlin JH (1984) Phys. Med. Biol. Vol. 29, No. 9: 1011
20. Vourvopoulos G, Womble PC (1989) Nucl. Instr. Meth. B36: 200
21. Kitto ME, Anderson DL (1985) Trans. Am. Nucl. Chem. Soc. 151
22. Zeisler R, Stone SF, Sanders RW (1988) Anal. Chem. 60: 2760
23. Rossbach M (1991) Anal. Chem. 63: 2156
24. Certificate of Analysis, SRM 1572, citrus leaves, National Bureau of Standards, Washington D.C. 1982
25. Certificate of Analysis, SRM 1575, pine needles, NBS, Washington D.C. 1984
26. Certificate of Analysis, SRM 1549, Non-fat milk powder, NBS Washington D.C. 1984
27. Certificate of Analysis, NIES Nr. 8, vehicle exhaust particulates, National Institute of Environmental Studies NIES, Japan 1989
28. Certificate of Analysis SRM 1577a, bovine liver, NBS Gaithersburg, M.D. 1982
29. Report of Analysis, RM 8431a, mixed diet, National Institute of Standards and Technology, Gaithersburg M.D. 1989
30. Gladney ES (1980) Analytica Chimica Acta 2 385
31. Gladney ES, Burns CE, Perrin DR, Roelands J, Gills TE (1984) NBS Spec. Publ. 260
32. Cortes-Torro E, Parr RM, Clements SA (1990) IAEA/RL/128 (Rev. 1) Vienna 1990
33. Nielsen FH (1990) New essential trace elements for the life sciences. In: Zeisler R, Guinn VP (eds.) Nucl. Anal. Meth. in the Life Sci. Humana Press, Clifton N.J. 599
34. Wagner G: .Richtlinie zur Probenahme und Probenbearbeitung, Fichtentriebe (*Picea abies*). Fassung 4, April 1990. unpublished
35. Schladot JD, Backhaus F, Reuter U: Jül-Spez-330, Aug. 1985
36. Comar D, Cronzel C, Chasteland M, Riviere R, Kellersholm C (1969) The Use of Neutron Capture Gamma Radiations for the Analysis of Biological Samples. In: de Voe JR (ed) Proc. of the 1968 Int. conf. on Modern Trends in Activation Analysis, Vol. 1 NBS Spec-Pub. 312 p 114
37. Bowen HJM (1966) In: Trace Elements in Biochemistry, Academic, New York

Credit: This work was supported in part by the Ministerium für Umwelt, Naturschutz und Reaktorsicherheit, Bonn under the coordination of the Umweltbundesamt, Berlin.

Subject Index

The Handbook of
Environmental
Chemistry

Edited by O. Hutzinger

Volume 3

N. T. de Oude, Strombeek-Bever, Belgium (Ed.)

Detergents

Part F

1991. Approx. 460 pp. 48 figs. 116 tabs. Hardcover DM 248,–
ISBN 3-540-53797-X

Springer-Verlag
Berlin
Heidelberg
New York
London
Paris
Tokyo
Hong Kong
Barcelona
Budapest

Environmental Toxin Series

Edited by S. Safe, O. Hutzinger

Volume 3

S. Safe, Texas A&M University, College Station, TX;
O. Hutzinger, University of Bayreuth;
T. A. Hill, Washington, DC (Eds.)

Polychlorinated Dibenzodioxins and -furans (PCDDs/PCDFs)

Sources and Environmental Impact, Epidemiology, Mechanisms of Action, Health Risks

1990. IX, 145 pp. 8 figs. 32 tabs. Hardcover DM 174,–
ISBN 3-540-15552-X

Polychlorinated Dibenzodioxins and -furans (PCDDs and PCDFs) are potent environmental toxins. Environmental exposures to these compounds in part-per-billion (ppb) and part-per-trillion (ppt) concentrations are of particular interest. These exposures may arise from bleached paper products, paper production process sludge, effluent waste water from paper plants, consumption of food and water from contaminated sources and contact to paper products.

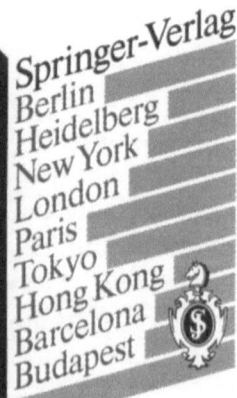

Springer-Verlag
Berlin
Heidelberg
New York
London
Paris
Tokyo
Hong Kong
Barcelona
Budapest